Essays in the History of Eugenics

Proceedings of a Conference organised by the Galton Institute, London, 1997

Conference organisers
John Timson, Milo Keynes and
John Peel

Edited by

Robert A Peel

PUBLISHED BY THE GALTON INSTITUTE

© The Galton Institute 1998

All rights reserved. No part of this publication may be reproduced or transmitted in any form or by any means without written permission or in accordance with the provisions of the Copyright, Designs and Patents Act 1988.

British Library Cataloguing in Publication Data

Essays in the history of eugenics
1. Eugenics 2. Eugenics - History
I. Peel, Robert A.
363.0'2
ISBN 0950406635

First published 1998 by The Galton Institute, 19 Northfields Prospect, Northfields, London SW18 1PE

Printed and bound in Great Britain by The Chameleon Press, 5-25 Burr Road, Wandsworth, London, SW18 4SG

Contents

Notes on the Contributors	v
Editor's Introduction Robert Peel	vii
The Theoretical Foundations of Eugenics Greta Jones	1
Eugenics: The Early Years Geoffrey Searle	20
Women, Feminism and Eugenics Lesley Hall	36
From Mainline to Reform Eugenics – Leonard Darwin and C P Blacker Richard Soloway	52
The Eugenics Society and the Development of Demography in Britain: The International Population Union, the British Population Society and the Population Investigation Committee Chris Langford	81
Human Genetics John Timson	112
Ninety Years of Psychometrics Paul Kline	128
The Galton Lecture: "The Eugenics Society and the Development of Biometry" Anthony Edwards	156
Eugenics in France and Scandinavia: Two Case Studies Alain Drouard	173
Eugenics in North America Daniel Kevles	208
Index	227

Notes on the Contributors

Professor Alain Drouard, Director of Research, National Centre for Scientific Research, Paris

Dr A W F Edwards, Reader in Biometry and Fellow of Gonville and Caius College, University of Cambridge

Dr Lesley A Hall, Senior Assistant Archivist, Contemporary Medical Archives Centre, The Wellcome Institute for the History of Medicine, London

Professor Greta Jones, Professor of History, Department of Humanities, University of Ulster

Professor Daniel J Kevles, Koepfli Professor of Humanities and Head of the Program in Science, Ethics and Public Policy, California Institute of Technology, Pasadena, California, USA

Professor Paul Kline, Professor of Psychometrics, Department of Psychology, University of Exeter

Dr Chris Langford, Reader in Demography, London School of Economics and Political Science

Robert A Peel, President, The Galton Institute, London

Professor G R Searle, Professor of English History, School of History, University of East Anglia

Professor Richard A Soloway, Eugen Merzbacher Chair, Department of History, The University of North Carolina at Chapel Hill, North Carolina, USA

Dr John Timson, Honorary Secretary, The Galton Institute, London

Editor's Introduction

Robert Peel

The annual conference of the Galton Institute has been a significant event in the academic calendar since 1963 when the first two-day symposium was held on the theme of "Biosocial Aspects of Social Problems". In each year the proceedings have been published in a variety of formats and to differing degrees of professional acclaim. This volume, comprising the papers presented at the Institute's 1997 conference, is the thirty-fourth in the series and it is the first to be devoted to eugenics.

The ninetieth anniversary of the founding of the Institute, as the Eugenics Education Society, was considered by Council an appropriate opportunity to review some aspects of its history and achievements. There is, of course, nothing inherently significant about a particular date - whether marking the end of a century or of a millennium - but nine decades represents a considerable life-span for a small learned society which, without either external funding or institutional backing[1], has depended on the financial support and voluntary efforts of its individual members and supporters.

That such an essentially private, self-financing and self-governing organisation, whose membership in its peak year (1932) barely exceeded 750[2] and whose philosophy ostensibly ran counter to the grain of fashionable thought during much of its existence, should have merely survived is alone surprising. That it can also be shown to have been influential in so many different fields is remarkable and probably unique.

The extent of this influence is attested by a considerable secondary literature on the Institute and its activities which has appeared during the last thirty years and which Council is presently cataloguing preparatory to producing a comprehensive bibliography. Scattered throughout the historical, educational, political and sociological journals and monographs, this extensive scholarly output describes the multi-directional efforts of the eugenics movement in its attempts to apply scientific ideas to the problems of society.

What those ideas were, where they originated and how they were fashioned into a philosophy and a programme by the founding members of the Eugenics Education Society form the subject matter of the first three chapters of this book.

Though deriving from a number of nineteenth century sources the philosophy of the eugenics movement was, as Greta Jones shows, uniquely a manifesto for an emerging meritocracy - though the word itself had not yet been invented. On its broadest interpretation it was nothing less than a theoretical vindication of the evolutionary transfer of social and political power from a landed oligarchy to a class which owed its growing influence solely to ability and achievement. A transition which, in France needed three revolutions, was accommodated in England slowly, silently and without upheaval and it was Galton's purpose to give legitimacy to this process by means of a theory which stressed its scientific inevitability, which linked it with the prevailing idea of progress but which also affirmed the value of existing structures, especially the family. Above all, Greta Jones argues, the theory derived support from the values of religious nonconformity. Thus, though Galton acknowledged his intellectual debt to Darwin he must also have been aware of a book which appeared in the same year in which the *Origin of Species* was published. This was Samuel Smiles' *Self-Help* which, selling a million copies during the author's lifetime and remaining in print to the present day[3], provided empirical support for the "survival of the fittest" in human society and gave the nonconformist work ethic its secular justification.

INTRODUCTION

In his *History of British Sociology 1834-1914*[4], Philip Abrams suggested that the history of eugenics "when it comes to be written will have to take account of political economy". Although Abrams' untimely death has deprived us of any further development of this idea it clearly relates to his observation, elsewhere in that book, that Galton's strictures on the poor and the destitute derived not from moral contempt but from what he saw as the sheer *economic* wastefulness involved in an inevitably growing underclass. Thus, in contemplating policies directed at this social sub-stratum, Galton was offering a programme which included negative, as well as positive, eugenics though by doing so he was abandoning, as Greta Jones points out, the tradition of Christian respect for weakness and contemporary ideas of equality.

But, though Galton had formulated the theory that inspired the movement that now perpetuates his name, he played no part either in the formation of the Eugenics Education Society or in its early development. The Honorary Presidency that he was persuaded to accept in 1908 was a plumage post and was abolished after his death in 1911. His own energies during his final years were spent in establishing the Galton Laboratory at University College, which received the whole of his considerable fortune at his death.

It is perhaps strange that Galton should have favoured the university department over the voluntary movement as a vehicle for his ideas. He had, after all, proposed eugenics as a religious creed and not only had he never held a university post, he had, at Cambridge, opted for a "poll" (aegrotat) degree. Geoffrey Searle's conclusion that he felt that a mass movement would coarsen his ideas is confirmed by Galton's niece who recalls his fear that the Eugenics Education Society would attract what he called "cranks"[5]. It is possible that Galton hoped for a close collaboration between the Laboratory and the Society but this was never achieved although for some years the Society helped to finance the Laboratory's *Annals of*

Eugenics. In the long term, however, the Galton Laboratory was more concerned with the development of techniques and methodologies than with the pursuit of a philosophical or social purpose. This latter task was left to the Eugenics Education Society, which was able to recruit its own independent scientific support when needed. The Society may well have had its fair share of cranks; as Geoffrey Searle shows, these were heavily outnumbered by a membership that included the most highly qualified scientists and doctors.

From its earliest days women have formed a significant proportion of the Society's membership. Three hundred and one out of a total membership of six hundred and twenty seven in 1914 were women. In the third chapter of this volume Lesley Hall draws on her familiarity with the eugenics archives and her expertise in the history of feminism to show why this should be so. The Duchess of Marlborough, Lady Ottoline Morrell and Lady Lutyens may have done no more than lend their names to what they regarded as a progressive cause; in similar organisations today they would be merely patrons. But there was a solid core of professional women who saw the Eugenics Education Society as a vehicle for their social aspirations. The six women members of Council in 1914 were representatives of this important group; so too was Sybil Gotto, "the virtual founder of the Eugenics Education Society" and its Honorary Secretary from 1907 to 1920. She was, indeed, the exemplar of the "New Woman", defined by Olive Schreiner[6] and fictionally immortalised by Wells as *Ann Veronica*[7]. It was the presence of this group within the Society that, as Lesley Hall observes, gave it its progressive, even leftish, orientation.

The shift from "mainline" to "reform" eugenics was the key event in the inter-war history of the Eugenics Society and, as Richard Soloway describes, was largely brought about by C P Blacker. This change of emphasis, together with Leonard Darwin's departure from the Presidency in 1929, enabled the Society to take a leading part in every organisational and scientific development in birth control in the next twenty years. Darwin had been reluctant to involve the Society in the birth control movement; he hated controversy (having perhaps seen enough in the parental home). But it was the Twitchin bequest, which he had secured for the Society, that gave it the means to provide material support for birth control and other endeavours. There is undoubted irony in the fact that the birth control movement which today regards eugenics with disdain was in the 1920s itself regarded as scarcely respectable by eugenicists. When, in 1974, a former British Health Minister (who whilst in office had been more generous than any of his predecessors to family planning programmes) gently suggested that these programmes should do more to target members of the Registrar General's Classes IV and V, the Family Planning Association hastily dissociated itself from any suggestion of "selective birth control". As Richard Soloway has commented: "Old time Eugenics Society members might well have wondered why they had invested so much in the birth control movement and the FPA before, during, and even after the war"[8].

The birth control movement originated as a neo-Malthusian activity; it subsequently derived theoretical support from the problem of differential fertility and in the 1930s operated against the background of the depopulationist panic. The difficulty of accommodating to changes of political and social climate was one amongst several influences that led to the demand for an adequate science of population. Chris Langford's painstaking analysis of the convoluted history of population studies in Britain reveals, once again, the influence of C P Blacker (and the resources of the Eugenics Society) in helping to establish demography as an independent scientific discipline. The Population Investigation Committee (a number of whose board members are members of the Galton Institute) remains a significant force in British demography and, through its journal *Population Studies,* an important influence internationally.

It was not merely a distaste for birth control[9] which prevented the Eugenics Education Society, in its early years, from developing a credible policy of negative eugenics; no theoretical basis for such a policy yet existed. Genetics, as a scientific specialism was, as John Timson describes, slow to develop in Britain following the rediscovery of Mendel's work in 1900 and it was only in the 1930s that *human* genetics became a serious branch of that discipline. From the start, however, human genetics developed within what Pauline Mazumdar has termed the "eugenics problematic", that is the agenda for negative eugenics established by the Eugenics Society. "The new human genetics of the thirties", she writes[10], "evolved in dialogue with the Society and its programme. The Society's interests formed the thread upon which the geneticist's pearls were strung." She points out that during the 1930s and 1940s there were few human geneticists who were not also members of the Eugenics Society. Human genetics is today the fastest growing branch of the life sciences and its achievements have brought it a glamour enjoyed by few other disciplines. They have also, as John Timson indicates, raised problems not very different from those that engaged the early eugenics movement[11].

Intelligence was a key concept in the eugenic ideology of merit. It was also the first human attribute to be measured by psychologists. In what must be the most succinct statement in the literature of sociology Michael Young expresses the relationship thus:

$$I Q + EFFORT = MERIT[12]$$

It is hardly surprising therefore that many of those pioneers of psychometrics referred to in Paul Kline's historical survey were closely associated with the eugenics movement. Sir Charles Spearman was an active member of the Eugenics

Education Society's Council from 1913 to 1919. Those writers who attempted, posthumously, to smear Sir Cyril Burt invariably referred to his connections with eugenics in a manner that suggested that this should rank amongst other offences to be taken into consideration. Adrian Wooldridge[13] has pointed to the debt which Ernest Jones, Cyril Burt and Raymond Cattell, amongst others, owed to Fisher in applying statistical techniques to their psychological studies. He also refers to the influential role of the *Eugenics Review* which published Burt's early work, one of the first articles on Binet's tests and the use of intelligence tests in the United States army in World War I. The Eugenics Society, he reminds us, was also responsible for the publication of one of the most controversial inter-war works of psychology, Cattell's *The Fight for Our National Intelligence*, written whilst Cattell held a Eugenics Society Fellowship.

Modern psychometrics, as Paul Kline shows, is concerned with personality factors other than intelligence which are likely to become of increasing significance with advances in behavioural genetics.

The Institute's Galton Lecture, given annually since 1914, is one of the high points of the annual conference. The 1997 lecture was given by Dr A W F Edwards and his subject was "The Eugenics Society and the Development of Biometry". C P Blacker defined biometry as "the scientific basis of eugenics"[14]. Yet to most members of the Institute, awed perhaps by the technicality of its methods and the complexity of its concepts, biometry has been seen as a somewhat esoteric, even forbidding, discipline. They will be grateful therefore to Anthony Edwards for his lucid account of the subject and of its relationship both to eugenics and to recent developments in neo-Darwinism.

At its foundation in 1907 the Eugenics Education Society was the first formal organisation based on eugenic philosophy.[15]

But during the early decades of the twentieth century similar bodies were formed worldwide. The societies which were set up in the countries of the former British Commonwealth were branches of the English Eugenics Society, often operating on the basis of its draft constitution, but elsewhere they were independent and each had its characteristic policies and practical programmes. Each, too, reflected the prevailing intellectual and political ideas of the country, and of the period, concerned. The last two chapters of this book, by Alain Drouard and Daniel Kevles, provide examples of these differences from Europe and North America respectively and furnish a valuable international perspective to the preceding essays.

The Eugenics Education Society avoided the unpleasant excesses of its North American counterparts; nor did it become involved with the issue of sterilisation as in Scandinavia. Instead, it sought to strengthen and clarify those scientific disciplines upon which its basic philosophy depended thus retaining its claim to scholarly status. The Society is perhaps an example of an organisation which failed in the task it set out to achieve but succeeded in other - and hardly less important - directions. Above all it clung consistently to its conviction that there is a biological dimension to individual behaviour and social affairs which merits scientific study. When even this basis proposition has come under challenge this may be seen as its most lasting achievement.

The Galton Institute is grateful to all those who gave papers at the 1997 Conference and thus contributed both to a highly successful meeting and to a book which we hope will attract a wide general readership.

Notes and References:

[1] It did not even enjoy the benefits of a registered charity until 1965.

[2] G R Searle, "Eugenics and Politics in Britain in the 1930s", *Annals of Science,* 36, 1979, p160.

[3] Samuel Smiles, *Self-Help,* Introduction by Sir Keith Joseph, Penguin Books, 1986.

[4] Philip Abrams, *History of British Sociology* 1834-1914, Chicago, 1968.

[5] Dorothy Middleton, *Sir Francis Galton 1822-1911,* Jubilee Memoir, Eugenics Society, London, 1982.

[6] Rosaleen Love, "Alice in Eugenics-Land: Feminism and Eugenics in the Scientific Careers of Alice Lee and Ethel Elderton", *Annals of Science,* 36, 1979, pp 145-158.

[7] H G Wells, *Ann Veronica* (1909). The phrase "New Woman" is said to have originated in an article in the *North American Review* in 1894. A more convincing portrayal is to be found in Grant Allen's *The Woman Who Did* (1895) but this work never achieved the popularity of Wells.

[8] Richard A Soloway, *Demography and Degeneration,* The University of North Carolina Press, 1990 and 1995, p 360.

[9] The Society's Council was divided on the eugenic effects of birth control; a substantial group held that since it was a predominantly middle class practice its effects were dysgenic.

[10] Pauline M H Mazumdar, *Eugenics, Human Genetics and Human Failings* Routledge, London 1992.

[11] See also John Postgate, "Eugenics Returns", *Galton Institute Newsletter* No. 21, 1996, p 4.

[12] Michael Young, *The Rise of the Meritocracy 1870-2033*, Penguin, 1961.

[13] Adrian Wooldridge, *Measuring the Mind,* C U P, 1994, pp 144-145.

[14] C P Blacker, *Eugenics: Galton and After,* Duckworth, 1952, p 105.

[15] It was also the last to change its name and the only one to survive.

Theoretical Foundations of Eugenics

Greta Jones

In 1865 Francis Galton (1822-1911) laid before the Victorian public in *Macmillan's Magazine* an article on *Heredity Talent and Character*. Four years later he published with Macmillan a full-length study of *Hereditary Genius*. In these publications Galton asserted two crucial facts; the first was the inequality of human beings - the fact that a few achieve distinction in life and most only modest or no distinction at all. Secondly, he attributed this inequality to heredity. Distinction did not always manifest itself in the same ways, but eminence in various professions and callings kept reappearing in the family histories he traced. Galton was also engaged in statistical studies, travel writings, meteorological observations and the study of individual human differences. The latter led to the publication of two large volumes *Inquiries into Human Faculty*, 1883 and *Natural Inheritance*, 1889.

The study of heredity, however, increasingly occupied his time. Galton believed that by measuring the differences and similarities between generations, it was possible to arrive at a statistical law of heredity. He tended to see this as a simple and commonsensical process - although horrendously complicated mathematically - by which each parent gives fifty percent of its inheritance to its immediate offspring, a quarter to its grandchildren and so on. In the 1890s his friendship with Karl Pearson, a philosopher and mathematician at University College, London, opened up access to the academic world. Pearson carried out studies of heredity along Galtonian lines at the Biometric Laboratory at University College and became the first professor of eugenics there in 1911 on the money donated by Galton in his will.

Galton and Pearson were wrong about the character of heredity. Mendelian theory, rediscovered at the turn of the century, eventually superseded biometrics. In Mendelism there is no infinite statistical division of heredity but a set of characters constantly jumbled up and redistributed in each generation. Nonetheless Galton's and Pearson's contributions to statistical methods were important in developing the mathematics of genetics. The Galton Laboratory, which evolved out of the Eugenics Record Office established at University College by Galton, was subsequently headed by a succession of the most distinguished pioneering geneticists in twentieth century Britain including R A Fisher, J B S Haldane and L S Penrose.[1]

Galton's work on identifying and measuring human individuals filled the pages of anthropological journals in the 1880s and 1890s. Anthropology in the nineteenth century - which included studies of European populations - was a measuring occupation and Galton's innovations in techniques, ideas and instruments were influential. The Anthropometric Laboratory in South Kensington that Galton set up in 1885 during the International Health Exhibition, was copied in Cambridge by John Venn in 1889 and by the Museum of Comparative Anatomy in Trinity College, Dublin in 1891. But the measuring of physical characters, whilst it remained a significant part of anthropology well into the twentieth century, declined in significance. From the point of view of posterity, the crucial contribution made by Galton was his investigations into physiological processes and mental and perceptual differences. The most politically and socially important outcomes of this line of investigation, relating closely to Galton's interest in intelligence, were tests of mental ability which, certainly in Britain, were undertaken in the spirit of Galton's eugenic creed.

THEORETICAL FOUNDATIONS

These contributions alone ensure Galton's importance in the history of studies of heredity, statistics and experimental psychology. But he also had a much more wide-ranging social influence. In his articles of the 1860s, Galton expressed the opinion that the amount of talent in society was limited but not necessarily finite. It was possible, he argued, to devise programmes which would identify the talented and encourage their fertility. By these means, society might be improved by increasing within it the proportion of the intelligent. Galton invented the name eugenics - used first, according to D W Forrest, in *Inquiries into Human Faculty* in 1883.[2] Galton's ideas were to have practical effect in the early 1900s, in the launch throughout Europe and North America, of societies for the implementation of practical eugenic policies including the Eugenics Education Society of London founded in 1907.

Where do Galton's ideas come from? Galton himself pointed out, there was evidence in the past of societies placing restrictions upon marriage which, it could be argued, were intended to have eugenic effects. There were other thinkers - for example Plato in the *Republic* - who advocated schemes for selective breeding to secure the public good. However, the eugenic philosophy launched in the 1860s by Galton had distinct historical characteristics. Moreover, what stands out is the extent to which eugenics in the early twentieth century caught the public imagination.

One reason was that Galton asked the question - also addressed in the 1860s in spate of books and articles - why societies progress. The unprecedented economic development and the relative political stability of the 1850s and 60s had led to European, and in particular British, dominance in the world. European and British intellectuals speculated about the factors that lay behind it.

The answer to the question was, frequently, the overriding significance of intellectual development. Henry Buckle in the *History of Civilisation in England* (1857) argued that climate and geography were factors which had forced societies in Northern Europe to devise the means to overcome cold and scarcity and, in the process, become more innovative. E B Tylor in *Researches into the Early History of Mankind* (1865) attributed animism and totemism in primitive societies to ignorance of the laws of nature. In Europe the progress of science led religion to become more philosophical and abstract and by implication sophisticated and progressive. John Stuart Mill was unusual in that he was not convinced that progress had occurred, certainly in morals or religion, but he exempted from this general rule intellectual development; 'Speculation, intellectual activity... is the main determining cause of social progress.'[3]   *On Liberty* in 1859 set out the means by which freedom of speech, thought and association might be safeguarded in modern societies and a major justification for their preservation was that intellectual progress would thereby be ensured. Bagehot in a series of articles in the late 1860s - later compiled and published as *Physics and Politics* (1872) - elaborated upon Mill's thesis.   Failure to develop liberal political institutions injured intellectual progress and was the cause of the static and unprogressive character of non-European societies.[4]

Galton's views on intelligence and progress were therefore widely shared among his contemporaries.   His exclusive concentration upon hereditary ability, however, was not. Bagehot argued for the role of liberal political institutions in human history; Buckle climate. In 1873, shortly before Galton published *English Men of Science* (1874), Alphonse Candolle in *Histoire des Sciences et des Savants depuis deux siecle* discussed the causes of scientific eminence, paying particular attention to its geographical distribution within Europe.   He attributed the lower levels of scientific achievement in southern European

THEORETICAL FOUNDATIONS 5

countries, to the influence of Catholicism. Scientific progress was made difficult in countries where a tradition of opposition between church and science existed and in which the financial resources of a nation were absorbed by the art and architecture of the Church. Mill went further and dismissed the notion of innate human characteristics as determining social development; 'the prevailing tendency to regard all the marked distinctions of human character as innate ... is one of the chief hindrances to the rational treatment of great social questions ...'[5]

Galton referred to Candolle's views and undoubtedly he admired the institutions of Victorian Britain as exemplifying excellence and progress. However, Galton began to resurrect the notion of the innate. This was not the philosophical justification for conservatism which would immediately be recognised and dismissed by the nineteenth century liberal - ideas and faculties planted by God in man and reflected in the permanence of traditions and ancient institutions and hierarchies. It was not institutions, hierarchies or religious precepts which were, in Galton's view, unchangeable nor even human nature but hereditary talent - identifiable through scientific investigation. So why did Galton's thoughts on the subject take the particular direction they did?

Hereditary Genius was a celebration of the rise to prominence of Galton's social class and family circle. In 1955 Noel Annan identified a group of families in the late eighteenth and early nineteenth centuries whom he called the intellectual aristocracy.[6] As opposed to the aristocracy of birth, they achieved a prominent position in British life primarily by their intellectual and professional labours. Galton belonged to this intellectual aristocracy. Annan described them as bound together by their religious origins in dissent. Although he was baptised into the Church of England, Galton's family were originally Quaker. The origin of their wealth was in industry or commerce or, occasionally, professional or legal labours.

Galton's family in the eighteenth century were Birmingham small arms manufacturers and later bankers. Marriage and family connections bound them together. Galton's family ties included the Wedgwoods, the Butlers, the Barclays and Darwins all identified by Annan as belonging to this group.[7] Above all, they were linked by certain common sentiments and opinions. They were liberals with a small 'l'; anti-slavery; advocates of the reform of government and parliament; pro-religious toleration. They were believers in laissez-faire political economy and free trade. At the same time they were moral and social reformers, the inspiration and mainstay of many philanthropic organisations of the early nineteenth century. They frequented and in fact helped create the cultural and intellectual organisations of urbanising Britain. At its most radical edges, the liberalism of early nineteenth century Britain lauded the accumulation of wealth by industrial entrepreneurship and professional work but denigrated land ownership. It attacked rent as an unjustified exaction upon enterprise and criticised entailment and primogeniture, two essential social instruments, which perpetuated aristocratic wealth and social position.

Lawrence and Jeanne Stone have pointed out how separate, until the late nineteenth century at least, the worlds of the rural gentry and the urban middle class were. The aristocratic families they examined over three centuries married among themselves and within their social circle.[8] So too did Annan's intellectual aristocracy. Davidoff and Hall in their book on the rise of the urban middle classes in provincial England, *Family Fortunes*, have described, for the period of the late eighteenth and early nineteenth centuries, how vital the institution of the family and the wider social links created by marriage and religious affiliation were for the success of the urban middle class.[9] The crucial position occupied by family provided Galton with two concepts - the idea of lineage and the idea of what constituted the social good.

THEORETICAL FOUNDATIONS

This was the milieu which nourished Galton and which makes *Hereditary Genius* a manifesto for this aspiring and ambitious social group. In it, Galton called for pride of race but not what he called 'the nonsensical sentiment in the present day which goes under that name.'[10] By nonsensical sentiment he meant the cachet attached to birth alone, the perpetuation of an ancient name, of old riches or the long association of a family with a place or estate - all of which would with, only a few exceptions, have excluded the social and family circles in which he moved. Galton celebrated eminence through achievement. In 1868 a Liberal Government was returned which went on to reform the army and civil service, the last bastions of aristocratic patronage, by introducing and extending competitive examinations and appointment on merit. *Hereditary Genius*, written before these events, excluded civil servants from the genealogies of talent. Galton 'did not take much notice of official rank ... except of the highest rank and in open professions.' The majority of members of parliament were still drawn from agricultural land-owning interests in mid nineteenth century Britain and therefore Galton excluded statesmen. Not mincing his words he claimed, that 'many men who have succeeded as statesmen would have been nobodies had they been born into a lower rank of life.'[11]

In *Hereditary Genius* Galton envisaged a meritocratic society in which social classes were in flux. It was a state where 'society was not costly; where incomes were chiefly derived from professional sources and not much through inheritance; where every lad had a chance of showing his abilities and, if highly gifted, was enabled to achieve a first class education and entrance into professional life by the liberal help of the exhibitions and scholarships he had obtained in his early youth; where marriage was held in high honour as in ancient Jewish times... where the weak could find refuge in celibate monasteries and sisterhoods and, lastly, where the better sorts of emigrants and refugees from other lands were invited and welcomed and their descendants naturalised.'[12] Even when, later in his life, Galton's liberalism was wearing thin, his background still exerted considerable pull over him. The parameters of his eugenic utopia in the unpublished novel *Kantsaywhere*, idealised the life of his grandparents and great grandparents - industrious, serious, frugal, religious, exemplifying the domestic virtues, concerned with the public good and of course fecund.

At the same time, this manifesto of mid nineteenth century liberalism and family values, was moving in new directions. Galton was fascinated by two things which came to him from continental theorists. One was the way in which social laws could be deduced from statistical regularities. This he read in the work of the French statistician Quetelet, whose influence in 1860s Britain, in the public health movement and elsewhere, was growing. Quetelet pointed to the remarkable consistencies which could be detected in apparently random and discrete facts. The second were the researches of Lavater, Gall and others into the relationship between mental and moral characters and human anatomy. Quetelet himself assumed that 'there is an intimate relation between the physical and the moral nature of man and the passions leave sensible traces on the instruments they put in continual action.'[13]

Beatrice Webb, who became acquainted with Galton during the 1880s, described him as exhibiting, 'three distinct processes of the intellect'. These were 'a continuous curiosity about, and rapid apprehension of individual facts, whether common or uncommon; the faculty for ingenious trains of reasoning; ... the capacity for correcting and verifying his own hypotheses by the statistical handling of masses of data whether by himself or supplied by other students of the problem.'[14] Galton hoped that statistical regularities in human conduct and the correlation of physical characteristics with mental and moral were the key to unlocking the laws of human nature. His interest in the mathematics of correlation derived from this. It opened up whole new fields of inquiry which, he believed, might lead to the discovery of causal connections between things. Galton's legacy was passed onto Pearson. The Galton Laboratory in 1911 was still conducting measurements of the cephalic and nasal index of groups of people and trying to establish correlations between these and other more imperceptible mental and moral characteristics.[15]

The second major influence was the publication of the *Origin of Species* in 1859. Shortly after its publication Galton entered into a correspondence with Darwin. The tenor of Galton's comments was, however, not altogether to Darwin's liking. Galton's opinion expressed in 1873 in *Fraser's Magazine* and conveyed to Darwin in the 1860s, that the struggle for existence 'seems to me to spoil and not to improve our breed ... On the contrary it is the classes of coarser organisation who seem on the whole the most favoured under this principle of selection and who survive to become the parents of the next', rather alarmed Darwin. Darwin was, in the 1860s, formulating a reply to his critics who attacked him on similar grounds for the inadequacy of natural selection to account for human and social evolution. Darwin's references to Galton in the *Descent of Man* in 1871 reflect his uneasiness, simultaneously conceding and denying Galton's belief that natural selection had failed in the case of modern society.[16]

Yet Galton's project and Darwin's *were* linked. They both attributed cosmic significance to heredity, even when they clashed, as they did in the 1870s, over the actual mechanism of inheritance.[17] For both, mating and the choice of a partner were central to evolution. For Darwin 'No excuse is needed for treating this subject in some detail for, as the German philosopher Schopenhauer remarks, "the final aim of all love intrigues, be they comic or tragic is really of more importance than all other ends in human life. What it all turns upon is nothing less than the composition of the next generation ... it is not the weal or woe of any one individual but that of the human race to come which is at stake.'"[18]

This was also Galton's view, although his perception was perhaps less nuanced than Darwin's. Darwin struggled with the theory of sexual selection as applied to humans. No simple formulaic statements about the question satisfied him. 'With mankind, especially with savages, many causes interfere with the action of sexual selection as far as bodily frame is concerned.' Mating involved, as Darwin saw it, physical attraction, mental ability, energy, wealth and social position. He pondered whether free choice in marriage actually existed in modern society, the extent of marriage across the classes and the different inclinations of men and women.[19] Galton's view of marriage by contrast remained straightforward. Marriage was a social and eugenic duty and responsible people, in whom the future of the race is invested, put 'love games' aside for the wider social good. It was left to Karl Pearson to consider the broader implication of eugenics for marriage and the relationship between the sexes.

Darwin and Galton also shared another preoccupation and this was with Malthusian theories of population. The issues raised in Malthus' *Essay on Population* in 1798 had become by the mid nineteenth century part of common intellectual currency. In political debate, Malthusianism was cited as conclusive proof of the impossibility of socialism. As T H Huxley put it in 1888 'unlimited multiplication', meant that 'no social organisation which has ever been developed, no fiddle faddling with the distribution of wealth' could produce a society free of poverty.[20] Clergymen cited Malthus in sermons about prudence and restraint. John Stuart Mill's Malthusianism led him, at one point in his life, to advocate laws against imprudent marriages.

For Darwin population pressure was the motor of competition within species and competition within species led to natural selection. Galton believed, however, that Malthusian checks to population growth - disease, war and famine - had failed in modern society. This meant that the characteristics in the individual which led to unchecked population growth – fecklessness, imprudence and lack of restraint – were flourishing. Mr Galton, Darwin wrote in the *Descent*, believed 'A most important obstacle in civilised countries to an increase in the number of men of a superior class was that the very poor and reckless who are often degraded by vice almost invariably marry early whilst the careful and frugal, who are generally otherwise virtuous, marry late in life so they may be able to support themselves and their children in comfort.'[21]

Galton's views upon what became known as 'differential fertility' between the social classes, were largely intuited. There was no modern social class analysis of fertility until 1911 and he formulated his theories before the fall in average family size among the middle classes, which became so prominent in the debate upon population in the 1900s. Simon Szreter has recently disputed the idea that there was a uniformly high birth rate among the working class in Victorian and Edwardian Britain. Patterns of fertility among the poor varied according to economic and social circumstances.[22] The widespread perception of uncontrolled and threatening fertility among the poor owed more to personal and political insecurities among the middle class in nineteenth century Britain.

So too did the picture of the imprudent and careless poor. In 1883 Beatrice Webb visited the Heyworths – her mother's family – in Bacup, Lancashire. The family had not – unlike the Potters, her paternal relations – risen to great wealth during the industrial revolution and were still artisans and mill hands. Her visit was, in part, a social experiment equivalent to her excursions into London's East End. She found the Heyworths hospitable, respectable, independent and religious. She concluded that 'mere philanthropists are apt to overlook the existence of an independent working class and when they talk sentimentally of the people they really mean the "ne'er do wells." It is almost a pity that the whole attention of this politician should be directed towards the latter class.'[23]

Nonetheless Galton was expressing the concerns, intellectual controversies and the values of his class and family background. Of even greater importance is that Galton is pivotal to understanding changes taking place within that stratum of society. His work is indicative of changes taking place in the liberalism and moral and religious values which had sustained his social and his family circle.

The generation of Darwin and Galton was different in many ways from that of their predecessors. Religious belief and the association with non-conformity had weakened and scientific certainties had replaced religious. Galton, robustly sceptical, nonetheless hoped eugenics might provide the basis for reverence for the family and commitment to the public good that conventional religion had supplied in the past. There was change in other respects. The money accumulated by the Darwin, Wedgwood and Galton families went 'cascading down the generations', to use a familiar phrase, and this enabled both Darwin and Galton to lead a life without financial or professional pressure. Yet, both felt the need to engage in work or to find a useful vocation and Darwin, in particular, was acutely aware of the disappointment which his failure to adopt a profession had caused his father. The gospel of work was still important for Darwin and he told Galton that 'excepting fools, men did not differ much in intellect, only in zeal and hard work.'[24]

The changes in Galton's outlook were more extensive. He was attuned to the growing mood of social pessimism in the second half of the nineteenth century. Galton doubted that progress was secure. The liberal agenda had begun to be criticised from within the ranks of liberalism itself. The dangers posed to property and laissez faire by the new democracy instituted by the Reform Act of 1867 worried many. Mill's effort to widen the franchise to include women by an amendment to the bill in 1866 threatened assumptions about female inequality. The controversy over slavery provoked by the American Civil War and the Governor Eyre controversy (1866) hardened attitudes towards race.

In 1873 FitzJames Stephen, brother of Leslie Stephen and both members of Annan's intellectual aristocracy, published an attack on John Stuart Mill's ideas of innate equality in a book *Liberty, Equality, Fraternity*. Distinctions of race, class and sex, which Mill had argued were largely socially generated and injurious to social progress, were, in FitzJames Stephen's opinion, innate, unavoidable and socially necessary. In Galton's view too, the social class, race and gender inequalities in Victorian society were the outcome of the ineluctable force of heredity. However, unlike FitzJames Stephen who called for an end to radical experimentation and change in society, Galton harnessed the old liberal reforming temperament to the new goal of altering the balance of social classes in society by the manipulation of heredity. FitzJames Stephen thought the reforming agenda was heading towards the destruction of all social distinctions. Galton offered a reforming agenda which held out the hope of preserving them. This was not the old religious morality of his ancestors or the political liberalism of his peers but it had elements of both. It wanted to preserve the family, encourage public spiritedness and do good. As far as liberalism was concerned it was satisfied with most of the changes that the nineteenth century had brought - it was not a sentimental conservatism but reforming, meritocratic and pro-science. What it abandoned was the Christian respect for weakness and Mill's belief in the equality of individuals.

Galton caught the mood of self-doubt and pessimism of the educated classes in the last quarter of the nineteenth century. A constituency for eugenics had emerged among many who, at that stage, had not heard of either Galton or eugenics. Britain was losing economic pre-eminence and her social institutions seemed increasingly fragile. Mid-nineteenth century liberalism had held out the hope of wide ownership of property, wealth filtering down through society, the economic emancipation of the working class through their own hard work and thrift and their acceptance of the duties and responsibilities of the property owning citizen. Instead, Victorian civilisation seemed to have produced a mass of urban poverty which combined in the 1880s with social disaffection.

Gareth Stedman Jones has pointed to the increasing resort to hereditarian explanations of poverty in the London of the 1880s in his book *Outcast London*. The phenomenon was by no means confined to the metropolis but can be traced in provincial towns and cities. In the forums where the educated urban middle class discussed the topics of the day - the Statistical and Social Enquiry and local Natural History and Philosophical Societies - the question of urban degeneration was raised in the 1880s, more than a decade before the outbreak of the Boer War in 1899.

In 1888 James Alexander Lindsay MD, later to become a member of the Eugenics Education Society, discussed in Belfast's Natural History and Philosophical Society two recent medical reports, by Sir Thomas Crawford, Director General of the Army Medical Department at the British Medical Association meeting in Dublin, and of Dr Milner Fothergill at the British Association meeting at Manchester. By a comparison of the heights of a sample of the population, including army recruits and town dwellers, over a period of time, they had arrived at the conclusion that town life led to the physical deterioration of its population. Conway Scott, executive Sanitary Officer for Belfast, argued in 1894, thirteen years before the foundation of the Eugenics Education Society, that 'Now is the time to form a great society, having for its object the attainment of the highest possible perfection of the human race, physically, intellectually and spiritually and such an association would be of the greatest value in correctly moulding public opinion, in guiding the action of Governments and generally promoting the attainment of the highest possible standards of national health.'[25] A year later in 1895 the Reverend Henry Osborne, in a letter to the Statistical and Social Enquiry Society of Ireland, argued that the state should institute compulsory physical examination of applicants for marriage and, in cases of physical or mental degeneracy, should have the right to forbid them. He believed that this would lead to 'a decided diminution of pauperism which is one of the most obstinate of our social problems. ... the qualities which make ordinary pauperism, such as improvidence, thriftlessness, self indulgence and the like, are all connected with imperfect organisation, physical and mental.'[26]

All this occurred in Belfast and it indicates that, by the 1880s there was in provincial cities a stratum of people concerned with the poor who believed that the social problems of the city were ineradicable by the means hitherto used. They included medical officers of health and those described in the syllabus drawn up at the London School of Economics for 1905-6 as likely to benefit from a course of sociological education. These were 'Councillors, Poor Law Guardians, Members of Committees of Philanthropic Institutions and Societies, Rent Collectors, District Visitors, Trade Union Officials, Scripture Readers, Workers in Settlements, Workshop or Factory Inspectors, Friendly Society Workers'.[27]

As the individuals and organisations dealing with the poor became more institutionalised and professionalised and as the hope, often religiously based, for moral transformation of the poor diminished, so the ideology of eugenics came increasingly to the fore. There was a cross fertilisation of hereditarian ideas and sociological concerns. In the 1900s the Charity Organization Society taught courses for social workers in conjunction with the Eugenics Education Society. J W Slaughter chairman of the Eugenics Society taught comparative psychology at the London School of Economics on the sociology course whose syllabus I have just quoted. E J Urwick, a member of the Eugenics Education Society, taught social administration there.

Eugenics became a social movement when it connected with a wider public opinion. There were dangers in this. Karl Pearson, for example, worried that the Eugenics Education Society would discredit itself by offering advice to the legislator without a sound theoretical understanding of heredity. By this, of course, he meant the biometric methods pioneered by Galton and himself. In fact neither biometrics nor Mendelism informed the early efforts of the Eugenic Society to devise practical programmes – only a vague sentiment about the importance of heredity. Even in the 1930s, when the geneticist Fisher assumed a leading role in the Society, this still applied to the larger number of its members. Some of the advice given by the Society upon mental deficiency was compromised by subsequent developments in Mendelian genetics.[28]

By opening up to the public there was also the possibility that eugenics would become prey to enthusiasts of varying descriptions. The implications of heredity for the family, the position of women in society, the use of birth control, for a programme of state legislation, for sexual morals were the subject of dispute. Wings and tendencies emerged. Tensions arose between Galton, Pearson and the members of the Eugenics Society.

However, eugenics was, eventually, to emerge with a particular programme, the task of defending society from the multiplication within it of the residuum of degenerate, unemployable and feckless. It did this by drawing a broad picture of the social consequences of unrestricted multiplication which, in the short run, was successful in mobilising considerable political support. In subsequent decades it became a weakness. Falling birth rates, the strategy of the political establishment in Britain to incorporate not alienate

THEORETICAL FOUNDATIONS

Labour, the loss of influence among some sections of the intelligentsia led to a new phase in the history of the eugenic idea.

Even so there were enduring theoretical bases to eugenics even if the interpretation of them changed at different historical periods; Darwin's notion of the cosmic significance of heredity; the importance of family but a more objective and distanced examination of its contribution to social good; the ideology of merit; the desire to reconstruct institutions along more rational lines; the importance of population and political economy; the idea of statistical regularity and the morally improving temperament of the British non conformist tradition. These, at the very least, were Galton's enduring legacy to eugenics.

Notes and References:

[1] The Eugenics Record Office was founded in 1905 by Galton's money. It was separate from Pearson's Biometric Laboratory. In 1906 it became the Galton Laboratory with Pearson as director.

[2] D W Forrest, *Francis Galton, The Life and Work of a Victorian Genius*, Elek, 1974, p. 162

[3] Mill, *A System of Logic*, 1843, vol. 2, p. 607

[4] The conviction that intellectual progress was the key to the emergence of European power extended to those societies most threatened by it. James R Pusey in his book *China and Charles Darwin*, Harvard, 1983, argues that the spread of Darwinian ideas among Chinese intellectuals was, in part, because he was seen as the iconic figure of European science and it was through scientific and technological superiority that the West threatened to swallow up China.

[5] Mill, *Autobiography*, 1873, p. 274

[6] Noel Annan, The Intellectual Aristocracy, in J H Plumb ed., *Studies in Social History*, Longmans, Green and Co., London, 1955, pp. 241-287

[7] They became known as the Clapham Sect

[8] Lawrence Stone, *An Open Elite? England 1540-1880*, Oxford University Press, 1984

[9] Leonora Davidoff and Catherine Hall, *Family Fortunes. Men and Women of the English Middle Class*, London, Hutchinson, 1987

[10] *Hereditary Genius*, 1869, pp. 46 and 9.

[11] The return of the Liberals in 1868 saw reforms in the civil service which instituted competitive examination for the administrative grades and in the army where the purchase of commissions was abolished. In addition, universities were opened to dissenters.

[12] Galton, *Hereditary Genius*, Macmillan, 1869, p. 362

[13] M A Quetelet, *A Treatise on Man*, 1842, p. 97

[14] Beatrice Webb, *My Apprenticeship*, Penguin, 1971 (1926), pp. 150-1

[15] Whilst this led to increasingly sophisticated statistics, it had by that time, become only marginal to the development of the science of heredity and when Fisher succeeded Pearson at the Galton Laboratory in 1933 he changed the direction of research. Nonetheless the *Treasury of Human Inheritance*, begun before the First World War, continued and did turn up significant pedigrees of vital importance in developing genetical science.

[16] See Darwin, *Descent of Man*, New York, Modern Library Edition, pp. 503-5

[17] Darwin's theory of pangenesis was put forward in a chapter in *Variation of Animals and Plants Under Domestication*, London, John Murray, 1868, and effectively disproved by Galton in a series of experiments in 1870-1. This also caused some friction between the two. see Forrest pp. 102-5

[18] Darwin, *Descent of Man*, 1871, p. 893

[19] 'Civilised men are largely attracted by the mental charms of women, by their wealth and by their social position, for men rarely marry into a much lower rank. The men who succeed in obtaining the more beautiful women will not have a better chance of leaving a long line of descendants than other men with plainer wives, save the few who bequeath their fortune according to primogeniture. With respect to the opposite form of selection, namely of the more attractive men by the women, although in civilised nations women have free or almost free choice, which is not the case with barbarous race, yet their choice is largely influenced by the social position and wealth of the men; and the success of the latter in life depends much on their intellectual powers and energy, or on the fruits of these in their forefathers.' *Ibid.*

[20] T H Huxley, The Struggle for Existence, *Nineteenth Century*, 1888.

[21] *Descent*, op. cit., p. 505. Darwin pointed out there were checks upon this particularly the high mortality rates among the very poor

[22] Simon Szreter, *Fertility, Class and Gender in Britain 1860-1940*, Cambridge 1996

[23] Beatrice Webb, *My Apprenticeship*, Penguin, 1971 (1926), p. 171

[24] Darwin to Galton, 23 December, 1870, in *More Letters of Charles Darwin*, Ed F Darwin, 1903, vol. 2, p. 41

[25] Conway Scott, National Health, *Belfast Natural History and Philosophical Society*, 1892-1900. (Paper given 3 January 1894), p. 38

[26] Henry Osborne, The Prevention and Elimination of Disease, Insanity, Drunkenness and Crime - A Suggestion, *Statistical and Social Enquiry Society of Ireland*, Part 2, 1895, p. 92

[27] London School of Economics Calendar 1905-6

[28] See, Greta Jones, *Social Hygiene in Twentieth Century Britain*, Croom Helm, 1986, pp. 97-8.

Eugenics: The Early Years

Geoffrey R Searle

Organised movements often put down roots long after the emergence of the ideologies that had inspired them. In Britain MPs were calling themselves "Liberals" some thirty years before the foundation of the Liberal Party, while "socialism" was a much discussed creed long before the first socialist societies came into existence in the 1880s. There was an equally long time lag with eugenics. Galton first used that word in his *Inquiries into Human Faculty* in 1883, and many of its leading ideas date back even further.[1] Yet the discussions to set up the Eugenics Education Society (hereafter the Society) did not take place until as late as 1907. Why its belated emergence as a public movement?

One reason is that, for many years after as well as before its establishment, there was uncertainty over what functions a eugenics society should serve: should it be the nucleus of a new political movement, or an ideas-based pressure group, or a centre of scientific enquiry?

Galton clearly inclined to the last of these options. Naturally he wanted to influence opinion-formers and policy-makers, but he never really reconciled himself to the prospect of having his ideas simplified and coarsened, as would inevitably happen once they were brought before a wider audience. His endowment of the Eugenics Record Office, renamed the Eugenics Laboratory in 1907, better exemplifies what he had in mind.

Even Galton's seminal papers to the Sociological Society, often taken as the start of eugenics as an organised movement, can be seen, if read carefully, to amount to little more than an appeal for promoting further scientific understanding: for example, "Eugenics: Its Definition, Scope and Aims" (delivered in 1904), advocates the "dissemination of a knowledge of the laws of heredity as far as they are surely known, and promotion of their further study", suggesting as useful projects an "historical inquiry" into differential fertility and a "systematic collection of facts showing the circumstances under which large and thriving families have most frequently originated". Galton also speculated about the possibility of eugenics later being "introduced into the national conscience, like a new religion", but he ended by warning against "over-zeal leading to hasty action".[2] In two papers given the following year, Galton again spoke of the way in which religion and custom might change human attitudes towards procreation, even envisaging that "in some future time, dependent on circumstances", "a suitable authority" might issue Eugenic certificates to those who applied for them.[3] But he said nothing about what "a suitable authority" might be: on the question of possible political strategies, Galton remained almost entirely silent.

So, not surprisingly, when a band of enthusiasts clubbed together to form the Eugenics Education Society, Galton initially held aloof, and it needed the personal entreaty of a long-time friend and neighbour, the lawyer, Montague Hughes Crackanthorpe, to persuade him to give the venture his blessing by accepting the office of Honorary President - a reluctant acquiescence which he later had cause to regret.[4]

In its early days the Society largely fell under the control of the proselytisers, among them its first chairman, an American sociologist bearing the unfortunate name of Dr Slaughter. But eugenics' most tireless advocate at this time was the physician and medical writer, Dr Caleb Williams Saleeby, who proclaimed in 1909 that the new creed was "going to save the world". However, extravagant language of this sort, allied to a bumptious manner, caused widespread offence; besides which, Saleeby was constantly annoying his colleagues by linking eugenics to other controversial causes, notably temperance. In

1910 he was voted off the Council, and in 1913 his offer to read it a paper was formally rejected. Saleeby thereupon became an outspoken critic, even an enemy, of the Society, attacking what he called the "better dead" school of eugenists and complaining that the movement was becoming discredited through its association with reactionary class prejudices.

Mrs Sybil Gotto, the Secretary, seems to have been the unsung heroine who held the Society together in these troubled early years. Also, the replacement in May 1909 of the first President, Sir James Crichton-Browne, by the distinguished lawyer, Crackanthorpe, to some extent steadied the ship. But it is interesting that even Crackanthorpe's relationship with Galton came under strain - in part because Crackanthorpe, though himself an able mathematician and scholar, nevertheless saw that a movement setting out to recruit members could not observe the severe formalities of a university debating club, still less that of a learned society. In any case, Crackanthorpe, an elderly man (he was 77 years old when he became President) could be nothing other than a stopgap.[5]

Incidentally, the continuing tension between the Eugenics Laboratory and the Society contributed to the dispute between "purists" and "popularisers", the Laboratory speaking out for science in all its austere rigour and the Society protesting that some compromises needed to be made if the word were to be spread. This was in some ways a bogus dispute, since the Society could call to its aid quite as many scientific heavyweights as the Laboratory, while the very notion that the Laboratory's Director, the hot-tempered and belligerent Karl Pearson, was somehow above the hurly-burly of polemical strife is risible. But the issue at stake was real enough.

The Society only really established itself when Leonard Darwin became its President in 1911, a post he held until 1928. As the fourth son of Charles Darwin, Leonard's family name had a totemic significance, and he quickly succeeded in

THE EARLY YEARS

establishing his authority over the warring factions of which the Society was composed. But not even Darwin, of course, had any control over others who chose to speak in the name of eugenics.

Freelance devotees caused much damaging publicity. For example, in an address to the Society in 1910, George Bernard Shaw, in mischievous mood, came out in support of lethal chambers and free love - an effusion which intensely angered the Society's officers. The popular press was quick to pounce upon such entertaining "copy". Nor could anything be done with stories such as the one featuring a Hampstead resident who announced to the world that he intended to father a "Superman". The baby turned out to be a Superwoman and was christened Eugenette. The father was later prosecuted for keeping his flat in conditions of filth and neglect. This story, inevitably, received splash treatment in the mass circulation newspapers.

And herein lay the rub. Once eugenics emerged from the cloistered calm of the debating society and the laboratory and attempted to draw public attention to itself, it ran the risk of gross vulgarisation - particularly since all attempts at discussing the processes of human procreation too easily elicit the embarrassed snigger. This obviously weakened the Society's attempts to counter the misrepresentations of its many bitter opponents, of whom the Roman Catholics were particularly outspoken.

Nevertheless, under Darwin a resolution of sorts was reached on the question of the Society's role. Like many, perhaps most eugenists, Darwin was preoccupied with the issue of legitimate **authority**. Watching aghast the excesses of "democratic" politics, he wanted to assert the role of scientific expertise in public life - as a counter to the nefarious activities of demagogic politicians. Unlike Saleeby but like Galton, Darwin doubted the wisdom of "science" setting itself up as a political movement at all. As a result, the Eugenics Education Society soon dropped any aspirations it may once have had to build up a mass membership, and instead sought to combine the promotion of scientific research with the exercise of insider influence.

***

Nevertheless, although the Eugenics Education Society had a difficult birth, it did survive these early troubles. Its modest success I see as the outcome of the convergence of two quite different developments. First, the early twentieth century was a time when a number of important scientific break-throughs in the study of human heredity occurred. Galton's own work, carried on by his friend, biographer, and disciple, Pearson, lay principally in biometry, the origins of which can be traced back deep into the nineteenth century. But it was the re-discovery of Mendel's famous paper which made the crucial difference: an understanding of the principles of particulate inheritance promoted the rapid development of genetics, carrying with it the promise (the over-optimistic promise) that scientists were close to controlling or eliminating a wide range of undesirable human traits, both physical and behavioural. In the years just before 1914 important advances also took place in demography, psychometry, and in many branches of medicine.

Significantly many of the pioneers of these infant sciences rushed to attach themselves to the Society, receiving in turn generous coverage in the pages of the *Eugenics Review*.[6] Speculation about the motives of individuals is always hazardous, but it is probable that many scientific "experts", eager to promote their own disciplines and thereby their careers, saw an invaluable opportunity of impressing the world with the social and human benefits that would flow from the support and endowment of their chosen avocations. Meanwhile, eugenics could plausibly be presented as the public face of scientific progress - an important reason for what early successes it enjoyed.[7]

But there is a second explanation for the crystallisation of eugenic ideas in the public movement that quickly gathered pace before 1914. Mounting anxiety over the long-term implications of the differential birth rate played an important role. But the date of the Society's effective establishment, 1908,[8] was also significant because this was the year which saw the passing of old age pensions legislation, the framing of the People's Budget, and initial steps being taken towards the formulation of the National Insurance Act. The sudden appearance of welfare politics ended an unspoken agreement between the two main parties that they would not compete against one another for votes by promulgating rival schemes of tax-funded reform.

As I earlier argued in my book, *Eugenics and Politics*, the Liberals' breaking of this taboo and the rise of Lloyd George and Churchill, promoters of a new kind of social radicalism, presented those anxious to defend the *status quo* with a dilemma. A mere repetition of the stale adages of political economy no longer sufficed; nor had the old individualism emerged unscathed from the "national efficiency" movement earlier in the decade which had suggested that the modernisation of politics required a recognition of the creative potential of state action.

Herein lay the usefulness of eugenics. It, too, rejected the old individualism, putting in its place the rights of the collectivity, "the race". It, too, recognised the futility of attempting to regress to an unfettered market economy. But eugenics seemed, at the same time, to be offering a powerful critique of the prevailing school of social reform. Eugenists argued that human beings, having interfered so drastically with their own physical and social environment, must in future complement this by controlling their reproductive processes - thereby replacing natural selection by rational selection.

This line of argument often culminated in crude attacks on Lloyd George and Churchill, their promotion of welfare legislation being presented, not just as a waste of time, but as the source of positive harm - since, it was alleged, the "fit" were being subjected to punitive taxes to fund social programmes which promoted the "multiplication of the unfit", a development leading inexorably to race suicide. Indeed, eugenists attributed a wide range of behavioural problems and social conditions, including unemployment and pauperism, not to economic factors, but to genetic defect. This approach was powerfully represented within the Society by the likes of E J Lidbetter, whose absurd "pauper pedigrees" received respectful attention in the *Eugenics Review*.[9]

***

Of course, such uncompromising opponents of social reform did not have things all their own way. For, just as Social Darwinism (of which eugenics was an offshoot) could be adapted to fit almost every conceivable ideological stance, so eugenics accommodated itself to a wide range of ideological and political positions. Indeed, as Michael Freeden and others have argued,[10] eugenics, with its air of scientific authority, appealed to the "progressive mind" as well as to people of conservative disposition. Socialists, reforming Liberals, fighters for women's rights, advocates of sexual liberation: all could find in eugenics much with which they agreed. Some members of the Fabian Society, in particular, were drawn to a creed that they saw as undermining the old *laissez-faire* values and strengthening the authority of the expert.[11] It was certainly quite *logical* to view eugenics and social reform as complementary rather than as antagonistic creeds.

Neither was eugenics entirely suited to politically conservative positions, elevating, as it did, inherent "fitness", scientifically ascertained, over mere status considerations. Admittedly, it was possible to argue that class was a biological category and that racial fitness and social status now largely coincided as a result of generations of "sifting". But it was equally possible to argue the opposite: even to demand, as Shaw did, equality of incomes so as to widen to the maximum the field of sexual selection.

Another difficulty centred upon the issue of war. The pre-war Conservative Party, with its links to the landed aristocracy and gentry, was steeped in militaristic values. Yet warfare, especially if waged by armies organised on the voluntary principle, was dysgenic. This presented real problems for an enthusiast like the journalist, Arnold White, a life-long devotee of the National Service League who also tried to popularise the eugenics cause.[12] Eugenics in many ways cut across the rhetoric which celebrated country and Empire and thus stood in an uneasy relationship to the patriotism and nationalism which lay at the heart of early twentieth-century Conservatism.

Yet, in its institutional form, the eugenics that triumphed had a decidedly conservative hue.[13] Most official pronouncements, though allegedly non-partisan, exude a mistrust of progressive liberalism and a horror of socialism. From eugenical literature one gets the impression that Conservatives might be maddeningly dim but that they did not constitute the "enemy". Why did this particular strand of eugenics win out over the various brands of reform eugenics that also throve in early twentieth-century Britain?

One possible explanation may lie in the rise to eminence of Leonard Darwin. His predecessor, Crackanthorpe, had been a Liberal before breaking with his party over Irish Home Rule, and he never entirely lost his progressive sympathies: for example, even in old age Crackanthorpe vociferously opposed big armaments, sympathised with the cause of women's emancipation, and was a courageous advocate of birth control.

Darwin, on the other hand, had served as the Liberal Unionist MP for Lichfield between 1892 and 1895 (by this time Liberal Unionists were effectively Conservatives), and, before becoming the Society's President, he was active in the Unionist Free Food League, one of the most socially conservative of all political groupings. These experiences seem to have coloured his pronouncements as President of the Eugenics Education Society. True, Darwin held back from pro-war enthusiasm in 1914, while his natural tact and good sense led him to distance himself from the gross partisanship betrayed by anti-Liberal zealots like James Barr: he often vigorously denied that there was "any inherent and necessary conflict between heredity and environment, if they may thus be personified".[14] But Darwin showed none of the reforming ardour that characterised Crackanthorpe, let alone a man like Saleeby, and it is noticeable that his annual presidential addresses grew steadily more defensive in tone from 1911 onwards. During his long term of office, the Society, ostensibly neutral in its politics, moved more or less steadily to the Right. When a later generation, including C P Blacker, spoke of the importance of rescuing eugenics from the narrow, class-based version of that creed, they surely had Darwin at the back of their minds.

But there is a second explanation for why things developed in this way which owes nothing to personalities. Historians studying the history of Victorian pressure groups are broadly agreed that the chances of success were greatly enhanced when a movement concentrated on a single objective and refused to be distracted by "side issues": the classic case of this is the Anti-Corn Law League, whose main leader, Richard Cobden, wisely discouraged members from mixing up the crusade for free trade with other good liberal causes.

Arguably, the floating of a successful eugenics movement similarly depended on the popularising of the big idea that it embodied - if need be, at the expense of rival creeds. Saleeby may have been justified in believing alcohol (like syphilis) to

THE EARLY YEARS

be a racial poison and in seeing no contradiction between extolling healthy parenthood and improving the environment - for example, by providing clean air. But he threatened to blur the message by mixing up eugenics' big idea with extraneous matters. The same obviously applies to the attempts at integrating eugenics with socialism. Instead, the opposite stratagem was generally adopted - promoting the idea of genetic inheritance and ridiculing all other reform programmes as un-scientific nonsense. After all, as Darwin himself put it, *somebody* had to speak up on behalf of the unborn.

Once the Society had thus established its own distinct identity, it felt free to co-operate with other more conventional groups of social reformers when their interests intersected - as happened, for example, in the formulation of policies for controlling the feeble-minded or for dealing with the consequences of venereal disease. Indeed, in her study of "social hygiene" in twentieth-century Britain, Greta Jones has demonstrated a significant overlap of membership between the Society and pressure groups which adopted a more "environmentalist" approach to social problems but shared its preoccupation with disciplining the urban poor.[15]

***

What kind of person joined the Society and what social role, if any, was it performing in early twentieth-century Britain? On the eve of the Great War, the Society's membership stood at 634: a figure which underestimates its true support because it omits people from affiliated provincial branches, some of which, notably the Birmingham Heredity Society, were quite important in their own right. All the same, the eugenists formed a relatively small elite before 1914 - somewhat like the Fabian Society, which in the first twenty years of its existence never had more than 900 people on its rolls. Things did not change much thereafter, the Great War dispersing the Eugenics

Education Society's membership, before a revival of interest took numbers up to an all-time peak of 768 in 1932-3.

From what backgrounds did these people come? Many commentators have noted the predominance of professional men and women, particularly doctors and scientists. Donald Mackenzie has gone still further, calling eugenics the "ideology of the professional middle class".[16]

Greta Jones, on the other hand, has forcefully challenged these prevailing assumptions.[17] First, she has shown that even the London branch of the Society contained a significant number of businessmen (at least in the 1930s) and that many of its professional members had important business **interests**: for example, Leonard Darwin, formerly an army engineer, held multiple company directorships. It thus makes little sense, she contends, to treat the professional middle class as though it formed a class by itself cut off from the wider capitalist society.

Second, Greta Jones suggests that it may be a mistake to concentrate, as most commentators have done, on London, which before 1914 was "largely a commercial and political centre with little large scale manufacturing industry", something that was "reflected in the composition of many of its organisations concerned with social policy and welfare" - for example, the Charity Organization Society, as well as the Eugenics Education Society. By contrast, she shows, powerful business families were well represented in the affiliated societies of Birmingham, Liverpool, and probably Manchester.

Professor Jones's third point is that the class character of eugenics should be defined primarily by its **function**; and, whatever the precise calling from which they drew their living, eugenists demonstrated their commitment to capitalism by their preoccupation with controlling and disciplining the urban poor - an obsession which made them very unpopular with nearly all working-class movements.

THE EARLY YEARS

These are all powerful and compelling arguments. But I am not entirely convinced by them. I concede that many, probably most, prominent members of the London-based Society had a stake in capitalism, whether as company directors or as shareholders. But while this explains their general conservatism, it does not dispose of the argument that the professional middle classes in early twentieth-century Britain were fashioning their own social perspective, one which emphasised expertise and an ethic of service rather than market success,[18] and that the eugenics movement epitomised this trend.

The point about London is well made. However, London may have been anomalous. The Edinburgh branch, for example, was set up by local "medics", following Darwin's address to its Medico-Chirurgical and Obstetrical Society, while at Oxford the running was predictably made by dons. In any case, it seems perfectly reasonable for historians to focus their attention on the parent society in London, which exercised a general control over the provincial branches, published the *Eugenics Review*, organised deputations to government departments, and so on - it is surely here, if anywhere, that one must look when trying to estimate the national impact of the eugenics movement.

That professional men were over-represented and businessmen under-represented in the Society may not in itself be that significant - the same is true of most pressure groups, and for understandable reasons. It **is** remarkable, however, that so little interest should have been taken by the Society in businessmen's concerns. For example, it was rare for anyone from a commercial organisation to address the Society,[19] and the *Review*, though it reported exhaustively on medical, scientific, and public health conferences, seldom if ever mentioned the social and economic anxieties expressed in gatherings like the National Association of Chambers of Commerce or later the Federation of British Industry.

The rhetoric of eugenics seems similarly detached from mainstream business concerns. In his 1913 Presidential Address, Darwin tried to estimate the "costs of degeneracy", but, though he did say something about how racial improvement would lead to enhanced industrial efficiency, his primary concern was to bemoan the fiscal burdens which social welfare had placed upon middle-class families, and middle-class families are, typically, presented as struggling, hard-working professional types. How significant, too, that when war broke out in 1914, the Society, Darwin to the fore, should have established the Professional Classes War Relief Council.

Most spokesmen for eugenics certainly viewed the urban poor with a mixture of fear and disgust, but this, by itself, did not make them enthusiastic capitalists. On the contrary, one constantly encounters an *anti*-capitalist strain in eugenical literature. "Passive capital" in the shape of an "idle" rentier class comes in for regular abuse for promoting an enervating luxury, while it is often insinuated that the quest for profit (though that word is seldom mentioned) was leading the business world to sacrifice the long-term future of the race - because, for example, businessmen benefited from the existence of a pool of cheap labour, even though that pool allegedly comprised many genetically defective human beings.

Moreover, when discussing desirable social types, eugenists nearly always focused on successful professional men, not entrepreneurs, a tradition dating back to Galton's *Hereditary Genius* of 1869.[20] This, in turn, I see as an expression of contempt for the "materialism" which eugenists held to be responsible for racial decay: for example, the desire to lead comfortable lives leading to the limitation of family size. Such a view of the world sits uneasily with capitalist values. The ethos of the Eugenics Education Society did indeed favour the pretensions of those whom Greta Jones has called "the dynasty of experts": but it evinced little sympathy for entrepreneurialism.[21]

On the other hand, I do not myself go along with Donald MacKenzie's view of eugenics as the "ideology of the professional middle class", if only because there are so many other contenders for that role which have equal or greater claims: for example, the Charity Organization Society, the Fabian Society, and various "New Liberal" coteries.

My own assessment would be that eugenics in early twentieth-century Britain, once seriously under-estimated, is now often accorded too much influence and significance, at the level of policy-making. Certainly there was a great deal of diffused eugenic thinking earlier this century, something, which both reflected and stimulated the emergence of various new scholarly approaches to the study of heredity. However, its impact on social policy was relatively slight: the Eugenics Education Society never succeeded in establishing for itself a position where it could articulate the wishes or interests of more than a small minority of either professional men or capitalists. Even more fatal to its chances of controlling the political agenda, the Society never recruited more than a tiny number of MPs.

At the practical level, the Society was most successful when it intervened on matters like the Mental Deficiency Acts - in other words, when it functioned as a kind of specialised pressure group which had the courage to speak out on controversial matters from which democratically accountable politicians prudently kept clear. This was a modest achievement, but not a negligible one.

More importantly, the Society has, in a variety of ways, some of them indirect, contributed to the advancement of an understanding of heredity and of population problems: genetics, biometry, and demography are the disciplines which have most obviously benefited from its support and encouragement. So perhaps, in the long run, it has been Francis Galton's own preferences which have triumphed, though they were once thought, even by many of his admirers, to be over-cautious.

References:

[1] Crackanthorpe published his *Population and Progress* in 1907, a collection of articles written over many years, the most important of which, "The Morality of Married Life", had first appeared in the *Fortnightly Review* way back in 1872. This is not the place to attempt a rounded portrayal of Galton's personality or of the development of his ideas. For recent interpretations, see Milo Keynes (ed.), *Sir Francis Galton, FRS: The Legacy of His Ideas* (London, 1993), especially the first two essays.

[2] *Sociological Papers, 1904* (1905), pp. 45-50.

[3] "Restrictions in Marriage" and "Studies in National Eugenics", *Sociological Papers, 1905* (1906), pp. 3-17.

[4] On the early history of the Society, I have drawn heavily upon my own *Eugenics and Politics in Britain, 1900-14* (Leyden, 1976).

[5] He resigned as President upon Galton's death in 1911.

[6] Thus volume 5 (1913-14) carries articles on psychology by William McDougall, on demography by W C Marshall, and on medical sociology by Edgar Schuster.

[7] Unfortunately the early eugenics movement became caught up in the war between biometry and genetics, each offering a seemingly different approach to the study of human heredity, and each viciously denigrating its rival. Broadly speaking, the Society quickly lined up behind the geneticists, while Galton's Laboratory and its Director, Pearson, took up the cudgels for biometry.

[8] Although the founding meeting was held in late 1907, the new Society only established itself during the following year.

[9] Later published as Heredity and the Social Problem Group (1933).

[10] M. Freeden, "Eugenics and Progressive Thought: A Study in Ideological Affinity", *Historical Journal*, 22 (1979), 645-71; D. Paul, "Eugenics and the Left", *Journal of the History of Ideas*, 45 (1984), 561-90.

[11] Though the commitment of leading Fabians to eugenics is often exaggerated (Searle, "Eugenics and Class", pp. 240-2).

[12] White was a member of the EES Council.

[13] See Greta Jones, "Eugenics and Social Policy Between the Wars", *Historical Journal*, 25 (1982), 717-28.

[14] Third Annual Report of EES, p. 7.

[15] G. Jones, *Social Hygiene in Twentieth Century Britain* (1986).

[16] See D. MacKenzie, *Statistics in Britain* (Edinburgh, 1981) and "Eugenics in Britain", *Social Studies of Science*, 6 (1976), 499-532.

[17] Jones, *Social Hygiene*, esp. pp. 19-21.

[18] The most powerful exposition of this view is Harold Perkin, *The Rise of Professional Society: England Since 1880* (1989).

[19] A rare exception was the address to the Society by the Secretary of the Scottish Life Assurance Company on the relationship between eugenics and National Insurance (Lewis Porr, "Insurance Research and Eugenics", *Eugenics Review*, 4 (1913), 331-55).

[20] Its largest sections are given over to judges and "divines". A few men responsible for notable "discoveries", like Watt and Brunel, get a passing mention in a chapter devoted to "Men of Science". Businessmen as such are ignored.

[21] On the other hand, there may have been a great congruence between eugenics and the corporate capitalism that was establishing itself during the course of the 1920s, as Greta Jones argues.

Women, Feminism and Eugenics

Lesley A Hall

It is often assumed that feminism and eugenics must be antipathetic, and certainly there is a long tradition of arguments about heredity and good breeding which have tended to regard women as simply vessels to receive the genetic contributions of worthwhile sires - that provided the sperm comes from a Nobel Prize-winner it matters less about the ovum. There is another view of the matter similarly denigratory to women, epitomised in the famous encounter, usually ascribed to George Bernard Shaw and Isadora Duncan, in which her looks and his brains are the desirable outcome, but the possibility of the reverse, quite the opposite! I don't think I have ever come across a version of this well-known 'urban myth' involving, say, a hunky Hollywood filmstar and a famous female intellectual.

However, as demonstrated in articles by my distinguished co-contributors Professors Soloway and Jones, in Lucy Bland's 1995 book *Banishing the Beast*, and a recent article by George Robb,[1] the relationship was in fact one of much greater complexity. Eugenics envisaged motherhood as being of central importance at the same historical moment as many women, both those committed to the feminist cause and those who were not, and even outright anti-feminists, believed in the revitalising of the nation through the improvement of the conditions and status of motherhood. Insofar as eugenics was in harmony with this, it was a cause to which women were inclined to be sympathetic, even if they did not regard heredity as the only problem in maternal and child welfare. I shall explore the relationship between eugenics and feminism between the wars, firstly, by looking at the uses made of eugenic concepts by specific women when talking about motherhood, health, and women's place and role within society generally, and secondly, by considering the relationship between the Eugenics Society and various women's organisations and campaigns. I should like to emphasise that this is far from the last word that can be said on the subject, and that I hope to draw attention to some useful areas for further study and analysis.

This was a period at which advice manuals and women's magazines were informing middle-class women, and health visitors were telling women of the lower classes, how they might actively promote their own health and that of their children. The approach to motherhood was becoming a matter of intervention and conscious choices, rather than resignation before the powers of fate and tradition, whereas eugenics tended to place women in a rather passive position. They were either good, fit, stock, in which case (provided that they were married) they were supposed to have as many children as they could, to replenish the nation, or, if they were of unsound stock, they were to refrain from breeding.

Some women were ardent eugenists who saw themselves as the educators of their own sex on a thoroughly 'them and us' model. Lady Barrett was an eminent woman doctor in gynaecological practice, on the Eugenics Society Council, and a leading figure within the Medical Women's Federation. She was an archetypal example of the sort of woman who joined the Eugenics Society, a highly qualified professional with interests in women's and children's health. She was over 40 when she finally married and there were no children of the union. This did not stop Barrett from arguing that she 'would rather encourage the fit to bring up families and restrict the unfit - not by propaganda, because that would not touch them - but by state interference', although she did consider that 'The question ... involved problems of housing and the conditions generally of workers'.[2] She advised women doctors that in 'private consultation ... every effort should be made to persuade normal healthy individuals to abandon the practice' of birth control, rather than acceding to demands for information.³ Her roots in the turn of the century social purity discourse peep out in her assertion that 'the suggestion that a knowledge of birth control would cure all their trouble is to deliberately hide the real thing, which is the unreasonable demands of their husbands'.⁴

Arabella Kenealy, a passionate anti-feminist, had also trained as a doctor during the 1890s but retired from practice due to ill health, becoming a novelist and writer on social reform. In her 1920 volume *Feminism and Sex-Extinction* she claimed that 'Nature made women ministrants of Love and Life, for the creation of an ever more healthful and efficient, a nobler and more joyous Humanity. Feminism degrades them to the status of industrial mechanisms.'⁵ 'True motherhood' she argued, 'is the greatest of the Creative Arts; Mother-craft the most vital and complex of the Sciences'.⁶ Since 'her momentous function of motherhood empowers her to make or to mar the Race', the nation 'has a greater claim upon its women, and has, at the same time, more reason and more right to restrict their liberty of action'. Kenealy conceded that 'To compensate her ... obviously [the nation] owes her privileges, personal and economic',⁷ and she even advocated a separate parliamentary chamber of women to foster the interests of women and children.⁸

Although associated with the Eugenics Education Society in its early years, and praising it as 'that admirable institution', Kenealy considered that its 'propaganda has been too much in the direction of eliminating defect from the Race by prohibiting marriage to the so-called "Unfit"'. Might it not be the case, she suggested, that as 'we are not in the secret of Nature's aims, and are wholly in the dark as to the human type for which she is aiming, to prohibit parenthood to any but the flagrantly abnormal ... might be to quench the evolution of such higher

Fitness as we are not qualified to foresee. That which shows like disability in one age may be the incipient ability of a later.'[9] She was therefore strongly opposed to what was usually a central tenet in any eugenic programme, sterilisation, believing that 'summarily to extinguish any human strain, by arbitrary prohibition, would be to exterminate a unique branch of the great Life-tree, and thereby to deprive the Race of a specialised route of further ascent ... Nature ... can judge as to what is intrinsic Fitness for Survival'. Thus, 'The doctrine of operative sterilisation is not only humanly repugnant but, in view of the psychological import of every physical function, it is essentially evil'.[10] Anti-feminist though she was, Kenealy retained a role for women of social rather than individual motherhood: 'to mother, befriend and inspire humanity at large rather than to minister to individuals ... to extend the tender, purifying ethics of Woman and The Home ever further and more deeply into public life'.[11]

A very different set of arguments was propounded in Marie Stopes's *Radiant Motherhood*, also published in 1920. This volume forms, as it were, part three of a trilogy with *Married Love* and *Wise Parenthood*. While hymning the joys of happy healthy motherhood and making recommendations as to how this was to be achieved, Stopes, formerly an active suffragette, commented that 'the best woman' was the one who 'out of a long, healthy and vitally active life, is called upon to spend but a comparatively small proportion of her years in an *exclusive* subservience to motherhood'. She had no truck with the argument, central to Kenealy's doctrines, that the normal woman should not 'exploit her capacities for individual gain, but for the benefit of her descendants'. Stopes characterised this as 'an endless chain of fruitless lives all looking ever to some supreme future consummation which never materializes.' To her way of thinking, this 'perpetual sinking of woman's personality' was 'a mistaken interpretation of her duty to the race'. Using 'intellect for individual gain in creative work' was not only of value to the community but would, according to Stopes, make women better mothers as well.[12]

Stopes has often been identified as a hard-line eugenist, drawing a clear distinction between the fit and the unfit and concerned to forcibly prevent the latter from breeding. Certainly in *Radiant Motherhood* she made a passionate plea for legislation to enable the sterilisation of the 'hopelessly rotten and racially diseased', claiming that these 'would be the first to be thankful for the escape such legislation would offer from the wretchedness entailed not only on their offspring but on themselves'.[13] But although this is the kind of statement for which she is remembered, *Radiant Motherhood* in fact advanced rather different arguments for the improvement of 'the race'. According to Stopes 'Baby's right to be *wanted* is an individual right which is of racial importance.' What she described as 'The physical and mental aberrations which are today so prevalent', she attributed to the prenatal effects of 'reluctant, perhaps horror-stricken, mothers', whose 'secret revolt and bitterness' generated a 'starved and stunted outlook' in the brains and bodies of their children.[14] Stopes had a powerful belief in the importance of pre-natal influence, which she suggested was mediated through the effects of the recently-discovered internal secretions.[15]

Though considerably less conventional in her views on motherhood than Kenealy, Stopes still assumed that happy healthy motherhood would take place within conventional monogamous marriage. The extreme left-wing feminist Stella Browne was one of the few overtly articulating a case for the unmarried mother, suggesting that many potentially good mothers were being denied maternity for want of a husband. Browne even slyly suggested that the eugenically desirable man possibly ought to make himself available for stud service. But I would argue that Stella Browne was deploying eugenic arguments to support her agenda of sexual liberation and women's choice, rather than promoting the latter for eugenic reasons. Her tart letters to *The Freewoman* and *The Clarion* on the class and gender biases of the Eugenics Education Society, and her allusion to its 'peculiar use of the terms "fit" and "unfit"' indicate that her vision of what it meant to be 'well-born' was at a wild tangent to that of the Eugenics Society.[16] Her views on sterilisation were, again, idiosyncratic. She constantly refused to countenance '*wholesale* sterilising or segregating', deploring the 'raucous hounding of the "unfit"', but also wondered 'why any sane and physically fine adult man or woman should not be able to be sterilised on demand' (i.e. as an efficient form of contraception).[17] Browne did, it is true, eventually join the Eugenics Society in 1938, but it seems probable that this formed part of the Abortion Law Reform Association's attempt to construct alliances to advance its aims. Her membership lapsed in 1942.[18]

These individual women, all of whom had a formal connection at some stage with the Eugenics Society, were nonetheless advancing very personal and often idiosyncratic arguments concerning eugenics and its implementation. In case anyone thinks that this is just a case of silly women who didn't really understand the subject and were wildly waffling on, similar idiosyncratic personal interpretations were common among large numbers of individuals of both sexes who were talking about eugenics at this period. But what the vast range of these were really talking about, and I doubt that it was a monolithic and internally consistent set of ideas, is rather beyond the remit of this paper. I will now proceed to the relationship between the Eugenics Society and women's organisations.

Following the achievement of suffrage, though not on fully equal terms until 1928, a feminist movement continued during the inter-war years in a more fragmented form. In particular specific campaigns, such as those for birth control and family allowances, have been characterised as embodying a 'New Feminism' based on women's special needs and responsibilities, which a discourse of political equality could not adequately encompass. These were areas in which both feminist and eugenic partisans took an interest, and within which strategic alliances might take place.

During the 1920s the Eugenics Society could not have helped but be aware that there were a number of vigorous women's organisations with national headquarters and local branches passing resolutions and lobbying for various causes to do with women, the family, health issues, etc, as well as educating their membership in the duties of citizenship. There were a number of not entirely happy attempts to establish liaison with these bodies. In 1926 'lectures on ... Human Biology, or Biology of Reproduction' were offered to the National Union of Women Teachers, whose response does not survive.[19] In February 1928 the Eugenics Society suggested to the National Union of Societies for Equal Citizenship (previously the National Union of Women's Suffrage Societies, the non-militant organisation), apropos of its conference on the training of unemployed women, 'There is one side of women's employment... important eugenically and that is domestic service.... lack of the right type of domestic help is, to some extent, responsible for the very great limitation of the families of educated and responsible people'. NUSEC responded politely that they had 'no scheme bearing directly on this question'.[20] In the same year the Society was itself approached by the National Society for Women's Service, requesting support for a deputation to the Chancellor of the Exchequer about the inequities affecting women within the Civil Service, particularly 'the difficulty of marriage for intelligent young women who have embarked upon a profession [which] does undoubtedly affect the low birth rate in the professional classes'.[21] The dysgenic effect of the marriage bar was also argued by Elizabeth Wilks in a pamphlet issued by the Medical Women's Federation in 1923: such women, she asserted, 'generally belong to a class selected for good health, good morals and good mental power...

particularly likely to hand on a good heredity.[22] The Eugenics Society, however, does not seem to have taken this particular issue on board and ingratiated itself with these powerful female interests.

There has often been asserted to have been a strong connection between eugenics and the birth control movement. By 1925 a number of birth control groups in the UK were setting up and running clinics and agitating for changes in government policy, and were largely run by women, many of whom had been active in the suffrage movement or influenced by it. The Eugenics Society, however, persisted in its pre-war attitude towards birth control as dysgenic since it was practised by the more responsible in society rather than those whom the Society thought should be restricting their reproduction. However, by the mid-20s an interest was conceded in methods of contraception which might be suitable for the least desirable members of society, and in 1926 the Society circularised birth control clinics for information on their clients, including the occupations of husbands (presumably to evaluate exactly what groups were employing contraception).[23]

In 1927 a Birth Control Investigation Committee was set up to undertake research in contraception in all its aspects, collecting statistics and evaluating methods. Largely at the urging of C P Blacker, the Eugenics Society supported this body with funding, and several members were represented on it. The BCIC became involved in analysing and testing existing methods of contraception and in developing what it was hoped would be more reliable, safer, and easier to use methods. It was unusual among birth control organisations in being predominantly male. In 1930 existing birth control organisations (including the BCIC) came together to form a National Birth Control Council, later Association. The Eugenics Society as such did not formally affiliate, although there were strong informal links through individuals active in both bodies and via the BCIC.

A facility which the Eugenics Society made some attempts to offer to prospective mothers was what would now be called genetic counselling. A letter of 6 October 1930 to Eva Hubback of the Society in her capacity as President of the Hampstead Heath Babies Club, suggested that 'some parents might be eager to get advice on the heredity side of their family problems. I wonder whether your committee would consider the possibility of co-operating with us in some way.'[24] This was apparently part of a larger 'outreach project': but it is not clear whether this initiative got off the ground, though the Society's records do include correspondence with members of the general public seeking advice as to whether they should have children.

Another area in which there was an overlap of interest between the Eugenics Society and women's organisations was the agitation for legalisation of voluntary sterilisation during the early 1930s. This has been masterfully analysed by John Macnicol, who points out that women's organisations displayed considerable interest in the subject, even left-wing bodies such as the National Conference of Labour Women and the Women's Co-operative Guild, although eugenics and sterilisation were generally anathema within the Labour movement. He comments that these women 'made an intuitive but confused connection between voluntary sterilization and broader issues of maternity'.[25]

It would initially appear that Blacker, on behalf of the Society, was making most of the running: writing on 27 February 1931 to Eva Hubback urging her to push the issue within the National Union of Societies for Equal Citizenship and the National Council of Women;[26] approaching the woman doctor Doris Odlum on 5 April 1932 about moving a resolution on eugenic sterilisation at the Women's National Liberal Federation Annual Conference;[27] and on the 29 of the same month asking Lady Denman of the NBCA if there was any chance of the Women's Institutes passing such a resolution.[28]

However it is clear that the matter was one of considerable concern to women. Detailed reports returned by the Society's lecturers on meetings they addressed throughout the country reveal that there was a groundswell of support among women for legalising sterilisation. This had diverse roots and was not simply brought about by the contemporary 'moral panic' over the alleged increasingly high proportion of 'defectives' within the population. Complaints were recorded along the lines of 'why the doctors didn't get the Bill passed since it was naturally for the good of the country',[29] 'capitalists who prevented the legalisation of any measures that would principally benefit the working classes',[30] and in some places 'People seemed surprised that parents of a defective child cannot now have the child sterilised'.[31]

There was some confusion about the meaning and effect of the operation: questions were asked such as 'Couldn't you apply compulsion in cases of persistent incest?',[32] or 'would a man convicted of assault on children be "cured" by sterilisation'.[33] Other women expressed anxieties around being 'told by a doctor that ster[ilisation] meant no more periods, no more sex pleasure', although 'One woman testified to married happiness after her own sterilization'.[34] Some women clearly saw sterilisation as a potential birth-control measure, and made occasional demands for a 'real voluntary measure [which] would enable anyone who wished to be sterilised', voicing queries as to 'why doctors in hospitals don't tell people - whom they think ought not to risk their lives with further pregnancies - that sterilisation is legally available now in therapeutic cases', and arguing to extend proposed legislation 'to poor women with already numerous children' or 'economic cases'. Others however expressed anxieties about 'mak[ing] it easier for the rich and idle to avoid having children' and 'normal people applying for selfish birth-control reasons?'.[35]

It was, however, frequently reported that 'there was a large group in favour of compulsion for MDs' and 'General feeling in favour of compulsion, as usual'.[36] There has recently been much journalistic horror about the views on eugenics and sterilisation of progressive intellectuals of the early twentieth century characterised as upper-middle-class élitists who despised the working classes. These audiences of lower class women indicate wide popular concern over these issues. Although the terms of debate appear to have been set by the contemporary 'moral panic' about mental deficiency, it was also strongly influenced by personal experiences and observations. Many of the women in these groups were reported as speaking from first-hand knowledge of the problems that mental deficiency could cause: what is not entirely clear is whether these women were concerned about the burden of a mentally-defective relative in families such as their own, or whether they were talking about what they perceived as an undesirable layer of the lower classes from which they wished to differentiate themselves. Both agendas seem to have been operating, as well as wider concerns around reproductive control.

The campaign for family allowances was a major 'New Feminist' issue. The Eugenics Society initially found the entire concept antipathetic, believing that they would simply encourage the feckless poor to breed even less cautiously: early Society campaigns to provide economic incentive for reproduction focused on alleviating the burdens on taxation on the middle-class household. By the 1930s Eva Hubback and C P Blacker in particular were arguing within the Society that family allowances were not necessarily dysgenic. In a letter of 13 June 1933 to R A Fisher, another member of the Society's subcommittee on the question, Blacker, while expressing some doubt that a scheme incorporating overt eugenic weighting could be practicable, suggested that family allowances would tend to encourage larger families among the kind of prudent thoughtful groups who were already successfully limiting their offspring.[37]

During the mid to late 1930s a rather uneasy courtship took place between the Eugenics Society and the National Birth Control Association, a body largely run by women and with, if not an overtly feminist agenda, one influenced by its members' background in other women's struggles. The Eugenics Society was a wealthy organisation, comparatively speaking, and the NBCA a poor one. By 1936 as well as funding the Birth Control Investigation Committee the Society was donating £100 a year to the Association's general activities. An informal meeting took place in December 1936 concerning 'closer relationship'. Though poor in cash, the NBCA was not lacking in attractive assets. In a memorandum of April 1937, Blacker suggested that the Executive Committee of the NBCA be constituted as a Birth Control Committee of the Eugenics Society, giving the Society access to the NBCA's nation-wide network of branches. He contrasted the success of the NBCA in establishing this network 'with the complete failure of the Eugenics Society to establish such branches', envisaging them becoming the 'working unit of a practical eugenic policy'.[38] In this courtship I am not sure whether the Eugenics Society can be regarded as a caddish and self-interested seducer or whether the NBCA was not, in fact, behaving like the gold-digger happy to accept mink and pearls but resenting any implication that she might be 'that kind of girl'.

On 28 April 1937 the NBCA Executive considered the report of the joint subcommittee on co-operation with the Eugenics Society. 'Several members expressed the opinion that some of these Aims and Objects [of the Eugenics Society] were highly controversial and that the whole project should receive fuller and further consideration'. At a subsequent meeting on 1 June, it was suggested that 'The present name, aims and objects of the Eugenics Society were unacceptable to the Association' and unlikely to find favour with the branches.[39] At a joint meeting on 8 July the NBCA stated that preconditions for amalgamation were 'a) education of the Branches and b) drastic alteration in the Aims and Objects of the Eugenics Society' and that in their opinion 'such amalgamation would be assisted by a change in the Eugenics Society's present name'.[40] So matters had proceeded from the Eugenics Society incorporating the NBCA to a demand that the Society itself change its name. By 28 September the Executive Committee of the Eugenics Society reported that 'conversations ... with the NBCA had for the time being broken down because of the reluctance of the Association to ask their branches to give unconditional approval to the Society's Aims and Objects', as a result of which a new version of these was drafted.[41]

In spite of this discord, in February 1938 the NBCA moved into premises in the Eugenics Society's building at 69 Eccleston Square.[42] However, this apparently cosy and intimate menage did not lead to any formalisation of the relationship. In the following year, on the NBCA's becoming the Family Planning Association, its objects led off with the classically Malthusian statement 'to advocate and promote the provision of facilities for scientific contraception so that married people may space or limit their families and thus mitigate the evils of ill-health and poverty', with no mention at all of eugenics.[43]

I should like to conclude by suggesting that the common ground upon which eugenics, feminism, and women in general could come together was motherhood and child welfare. Apart from those women who saw themselves as educators destined to preaching the eugenic gospel to other women, the appeal of eugenics to women was through the possibilities it offered, or seemed to offer, for assisting women's desire to bear and rear healthy wanted children when they chose to do so, in fact when it could be used to increase their choices. Insofar as it fitted in with this, there were possibilities of constructing alliances, but, as we have seen in the case of the flirtation of the NBCA with the Eugenics Society, these tended to be narrow, specific, and contingent. The rise in recent years of the acceptability of pre-natal testing may similarly indicate that women are prepared to accept a form of eugenics within the context of a particular problem with potential impact on their own lives. There is an enormous difference between telling a woman that she shouldn't have children at all, which is pretty denigratory, and giving her the information to enable her to make a decision as to whether to carry a particular pregnancy to term, although this brings its own problems. So, there have been persistent areas of common interest to both feminism and eugenics; however, while their points of view may sometimes have overlapped, they have seldom been exactly the same.

Note and References:

[1] R. A. Soloway, 'Feminism, Fertility, and Eugenics in Victorian and Edwardian England', in S. Drescher et. al. (eds), *Political Symbolism in Modern Europe*, New Brunswick: Rutgers University Press, 1982, pp. 121-145; G. Jones, 'Women and Eugenics in Britain: the case of Mary Scharlieb, Elizabeth Sloan Chesser, and Stella Browne', *Annals of Science*, 51 (1995), pp. 481-502; L. Bland, *Banishing the Beast: English Feminism and Sexual Morality, 1885-1914*, London: Penguin, 1995; G. Robb, 'The Way of all Flesh: Degeneration, Eugenics, and the Gospel of Free Love', *Journal of the History of Sexuality*, 6 (1996), pp. 589-603.

[2] 'Medical Women and Public Heath Problems', *Medical Women's Federation Newsletter* Jul. 1922, p. 23; archive of the Medical Women's Federation in the Contemporary Medical Archives Centre, Wellcome Institute for the History of Medicine, CMAC: SA/MWF/B.2/1.

[3] Lady Barrett's Presidential Address, *Medical Women's Federation Newsletter* Nov. 1922 p. 15, CMAC: SA/MWF/B.2/1.

[4] 'Conception Control in relation to Eugenics and National Welfare', *Medical Women's Federation Newsletter* Jul. 1927, pp 19-26, CMAC: SA/MWF/B.2.4; Barrett's influence or lack of it on attitudes to birth control among women doctors is discussed in L. A. Hall, '"A suitable job for a woman"?: women doctors and birth control before 1950' in Larry Conrad and Anne Hardy (eds.) *Women and Modern Medicine*, Rodopi, forthcoming.

[5] A. Kenealy, *Feminism and Sex-Extinction*, London: T Fisher Unwin 1920, p. 218.

[6] Ibid, p. 123.

[7] Ibid, p. 252.

[8] Ibid, p. 269.

[9] Ibid, p. 89.

[10] Ibid, pp. 90-91.

[11] Ibid, p. 74.

[12] M. C. C. Stopes, *Radiant Motherhood: A Book for Those Who are Creating the Future*, London: G. P. Putnams' Sons Ltd, 1920, pp. 157-9.

[13] Ibid, pp. 223-4.

[14] Ibid, pp. 173-8.

[15] Ibid, pp. 137-42.

[16] L. A. Hall. '"I have never met the normal woman": Stella Browne and the politics of womanhood', *Women's History Review*, 6 (1997), pp. 157-182.

[17] F. W. Stella Browne, Review of Miriam Van Waters, *Where Girls Go Right*, in *The New Generation*, III, Jul. 1924 , p. 82; review of J. P Hinton and Josephina E Calcutt, *Sterilization: A Christian Approach*, in *Plan for World Order and Progress* (The journal of the Federation of Progressive Societies and Individuals), Vol. 2 no 10, Oct. 1935, p. 23.

[18] Eugenics Society archive in the Contemporary Medical Archives Centre at the Wellcome Institute for the History of Medicine: Minutes of Council Meeting of 14 Feb. 1938, CMAC: SA/EUG/L.10, and of 14 Apr. 1942, CMAC: SA/EUG/L.11.

[19] CMAC: SA/EUG/D.149 National Union of Women Teachers.

[20] CMAC: SA/EUG/D.147 National Union of Societies for Equal Citizenship.

[21] CMAC: SA/EUG/D.144 National Society for Women's Service.

[22] E. Wilks 'The Eugenic Aspect of Conception Control', CMAC: SA/MWF/B.4/4/3.

[23] Soloway, Demography and Degeneration, p. 188.

[24] CMAC: SA/EUG/C.177 Eva Hubback.

[25] J. Macnicol, 'Eugenics and the Campaign for Voluntary Sterilization in Britain Between the Wars', *Social History of Medicine*, 2 (1989), pp. 147-69.; 'The Voluntary Sterilization Campaign in Britain, 1918-1939', in J. Fout (ed.), *Forbidden History: The State, Society and the Regulation of Sexuality in Modern Europe: Essays from The Journal of the History of Sexuality*, University of Chicago Press, 1992, pp. 317-33.

[26] CMAC: SA/EUG/D.147 National Union of Societies for Equal Citizenship.

[27] CMAC: SA/EUG/C.256 Dr Doris Odlum.

[28] CMAC: SA/EUG/C.89 Lady Denman.

[29] CMAC: SA/EUG/D.245 Joint Committee on Voluntary Sterilisation, Report forms on meetings 1934-35

[30] CMAC: SA/EUG/D.246 Joint Committee on Voluntary Sterilisation, Report forms on meetings 1936 Jan-Apr.

[31] CMAC: SA/EUG/D.245 Joint Committee on Voluntary Sterilisation, Report forms on meetings 1934-35.

[32] CMAC: SA/EUG/D.246 Joint Committee on Voluntary Sterilisation, Report forms on meetings 1936 Jan-Apr.

[33] CMAC: SA/EUG/D.245 Joint Committee on Voluntary Sterilisation, Report forms on meetings 1934-35.

[34] CMAC: SA/EUG/D.246 Joint Committee on Voluntary Sterilisation, Report forms on meetings 1936 Jan-Apr.

[35] CMAC: SA/EUG/D.246 Joint Committee on Voluntary Sterilisation, Report forms on meetings 1936 Jan-Apr.

[36] CMAC: SA/EUG/D.246 Joint Committee on Voluntary Sterilisation, Report forms on meetings 1936 Jan-Apr.

[37] CMAC: SA/EUG/D.65 Family Allowances.

[38] CMAC: SA/EUG/D.18 Correspondence with the NBCA 1936-44.

[39] Family Planning Association archives in the Contemporary Medical Archives Centre at the Wellcome Institute for the History of Medicine: CMAC: SA/FPA/A.5/2 Executive and sub-committee minutes.

[40] CMAC: SA/EUG/D.23 Co-operation between the Eugenics Society and the NBCA 1936-7.

[41] CMAC: SA/EUG/D.18 Correspondence with the NBCA 1936-44.

[42] CMAC: SA/EUG/D.18 Correspondence with the NBCA 1936-44.

[43] CMAC: SA/EUG/D.20 NBCA: minutes, circulars etc 1937-9.

From Mainline To Reform Eugenics - Leonard Darwin And C P Blacker

Richard A. Soloway

Nearly half the ninety-year history of the Galton Institute - or the Eugenics Society as it was known until 1989 - was dominated by two people: Major Leonard Darwin, its president from 1911 to 1928, and Carlos P Blacker, General Secretary from 1931 until 1952. Men of different generations, they came to eugenics for different reasons, and often profoundly, if politely, disagreed about the directions the Society should take. Nevertheless, they formed a curious intergenerational alliance of moderates who frequently struggled to contain the fringes of a movement that attracted its share of embarrassing extremists.

The influence exerted by Darwin, the second youngest and longest surviving of the great naturalist's five sons, was more diplomatic, passive and accommodating than it was aggressively formative. Despite his own unwavering belief in the overwhelming predominance of nature over nurture, Darwin, during his long years as president, uncomfortably straddled the transition from orthodox or "mainline" eugenics, as Daniel Kevles has described it, to a more sophisticated "reform" or scientific eugenics, spearheaded by Blacker, that focused upon the social as well as the biological quality of the population.[1]

Educated at the Royal Military Academy, Woolwich, Darwin served for twenty years in the Royal Engineers as an instructor and as a member of several scientific expeditions, resigning with the rank of major at the age of forty on grounds of "uncertain health." His interest in eugenics had been aroused in the 1880s by its founder Francis Galton, his father's cousin, but Darwin eventually concluded that eugenic reform was not practical and his interests were diverted elsewhere. Turning to politics in 1892, he was elected as a Liberal Unionist member for Lichfield, his great-grandfather Erasmus Darwin's constituency, but was defeated in 1895 and never ran again.

For the next few years Major Darwin lived the life of a country gentleman, writing essays on bimetallism and municipal trade and serving, on Galton's recommendation, first on the Council of the Royal Geographical Society and then as its president from 1908 to 1911.[2] Yet aside from his reading and correspondence with Galton nearly two decades earlier, he had expressed little interest in eugenics until, in 1909, at the age of fifty-nine, he delivered a lecture to the new Eugenics Education Society. Two years later, much to his astonishment, as he was not even a member yet, he was asked to take on the presidency of the fledgling organisation, again, probably on Galton's recommendation. Until that point Darwin had regarded himself "as more or less of a failure." It was eugenics, he claimed, that suddenly gave his life purpose and meaning; he felt that at last he was doing work of great importance.[3]

Darwin's leadership of the Eugenics Education Society, as it was called until 1926, was essentially defensive and reactive. In the face of growing demands that the organisation accommodate itself to the nurtural benefits of birth control and child and maternal welfare, as well as to exciting breakthroughs in the new science of genetics, he struggled to preserve it as a selective, pronatalist propagandist agency, dedicated to encouraging the "eugenically fit" to have more children. Who constituted the biologically elect was, of course, a vexatious question that plagued the eugenics movement from its beginnings, but mainline advocates like Darwin only had to look at the successful, professional middle classes for the answer.

Blacker, by contrast, was the aggressive architect of a reform eugenics that focused on negative or restrictive policies, primarily birth control, taking into account the need to weigh more accurately the interaction between heredity and environment as it affected the qualitative reproduction of people in all classes. Despite frustrating disagreements, heated conflicts, occasional charges of treason, and indignant resignations, he patiently pushed and prodded the Eugenics Society for some twenty years to become a credible scientific and social scientific research organisation and an influential advocate of social investigation.

Initially the old guard, including Darwin, were not quite sure what to make of Blacker. An extraordinarily vigorous, wounded, decorated war hero who had turned to psychiatry as a result of his experiences in the trenches, he was already well-connected in medical and scientific circles when he became active in the Society in the mid-1920s. Unsure about Blacker's rather severe, aloof "personality" and his degree of dedication to what for some had become something of a religion, Darwin still had serious reservations when in 1930 the part-time position of General Secretary became available.[4] He told Sir Bernard Mallet, his handpicked successor as president, that Blacker was "not my ideal because I doubt if he would make it his life's job." Blacker was not prepared to "chuck" medicine or his work with the new voluntary birth control clinics established in the 1920s. Nor would he give up his position as medical secretary of the Birth Control Investigation Committee, which he had been instrumental in founding in 1927 "to investigate the sociological and medical principles of contraception."[5]

Unlike a number of orthodox mainliners, Darwin had grudgingly conceded that whatever his doubts about the value of birth control as a eugenic weapon, the Society had to reach some compromise with those of its members, like Blacker, who increasingly saw it as the most effective route to eugenic advancement. But Darwin wondered how the young psychiatrist would deal with those pioneers of the movement less tolerant of the direction eugenics was taking. He cautioned him that "during the last twenty years I have had to deal with tactless and difficult people all the time. It will be the same with you, and it will need great calmness and tact on your part. What you will have to do is to make the best possible [use] of much inferior material," and, as he described elsewhere, row with "a good many defective oars."[6] On more than one occasion in the years ahead the frustrated Blacker had to be persuaded from jumping ship.

It took all of the personal qualities Darwin had described for Blacker to turn the Eugenics Society from a confused, unfocused amateur propaganda organisation dabbling uncertainly in dubious science, into a quasi-professional foundation committed to family planning and the serious study of social and biological population problems. Darwin was never comfortable with the reform agenda that Blacker advanced, and initially thought that the new secretary would do no more than "fill a gap" for three years or so when the whole issue could be reconsidered. But when that time came he had to agree with the organisation's treasurer that Blacker had "done most excellent work and changed the Society from a bickering debating club into an institution really accomplishing something and getting a position of weight, even if you haven't been able to stop the bickering."[7]

However much Darwin was himself given to bickering about many of the new ventures reform eugenics embraced, he was never wilfully obstructionist and cultivated Blacker throughout the 1930s, often using his still considerable influence with the old guard to mollify some of the resentment directed at the general secretary and his "scientific" allies. Blacker, in return, was unfailingly consultative, regularly asked for Darwin's advice, and carefully explained initiatives that he knew were not to the retired president's liking. Though Darwin repeatedly apologised for bothering the busy Blacker with his long letters, essays and observations on eugenic matters, and routinely absolved Blacker from any obligation to respond, Blacker not only replied quickly but travelled down to Sussex to visit his ageing friend every three months or so throughout the 1930s. The visits only came to an end with the outbreak of the war.[8]

In his first presidential address in 1912, Leonard Darwin had described eugenics as "the practical application to social life" of his father's theory of evolution, complemented by Galton's goal of bringing "as many influences as can be reasonably employed, to cause the useful classes in the community to contribute *more* than their proportion to the next generation."[9] In an era of diminishing birth-rates, it appeared that the ablest, most successful, and presumably the most desirable classes were in fact contributing far less than their share to that critical proportion. Fears about the consequences of differential fertility, as I have argued at length elsewhere, proved to be a major stimulus to the new eugenics movement.[10] The aged Galton had come out of retirement to warn that "the possibility of improving the race of a nation depends on the power of increasing the productivity of the best stock ..." rather than "repressing the productivity of the worst," and shortly before his death in 1911 he wrote to Darwin that "differential fertility ... to my mind is the most important of all factors in eugenics."[11]

Darwin interpreted such admonitions to mean that the role of the Eugenics Education Society was first and foremost to educate the better endowed representatives of current and future generations about the importance of marrying carefully and multiplying fruitfully rather than trying to induce the more fertile, but less well endowed, to limit the size of their families. In the early years of the century, anecdotal and comparative statistical ruminations nourished among eugenicists like Karl Pearson lamentations about the noticeable want of highly intelligent, talented people in science, the arts, commerce, politics and the professions once spawned by the able middle, and, sometimes, skilled working classes.[12] Although Pearson opposed the establishment of the Eugenics Education Society in 1907 as premature, his calculation that the most fertile and poorest quarter of the population - both economically and biologically - was producing more than half the next generation was regularly cited as exactly the reason why such a society was needed.[13]

At the time Darwin assumed the presidency, a growing minority of eugenicists were already questioning whether it was realistic to expect a statistically significant reversal of the diminishing birth rate among the eugenically desirable, and beginning to look more closely at the eugenic advantages of promoting birth control among the genetically suspect who allegedly populated the ranks of the poor. Mainliners like Darwin, however, were convinced that there was an intrinsic conflict between eugenics and family limitation because the personal qualities of prudence, foresight and self-restraint that birth control required were hereditary characteristics more common to the successful, educated classes than to the prolific labouring poor.

Before the outbreak of the war, Darwin had managed to keep the arguments for birth control from coming up for discussion in council meetings or in the pages of the Society's new journal, the *Eugenics Review*. He was helped by the knowledge that most of the men and women in the Eugenics Education Society still considered any public discussion of birth control distasteful. At the same time, he acknowledged to Havelock Ellis in 1917 that "possibly a bolder course might be better."[14]

What prompted this correspondence was a forthcoming article by Ellis in the *Eugenics Review* claiming that the war had put an end to "prudery and ignorance" so that birth control could now be adopted by eugenicists as "the magic formula ... to stem the tide of unfit babies." Darwin made sure that the article was prefaced with a note assuring readers that its publication did not mean a change in policy.[15] Given the extraordinary casualty rates decimating future fathers of all classes and an even more precipitous decline in the birth rate during the war, it was not a particularly propitious time to be advocating birth control.

Darwin publicly deplored the dysgenic consequences of fighting the war with a volunteer army and was an early supporter of conscription in the hope that the "casualty lists would then more nearly represent a random sample of the population" instead of a preponderance of those from whom it was most desirable that "the stock of the future" be produced. He was active in allying the Eugenics Education Society to The Professional Classes War Relief Council established in 1915 under his chairmanship to assure the safe delivery and care of children sired by officers of the professional middle class, "from which it is pre-eminently desirable that the largest number of children should be born."[16]

It was one of the few times that the Eugenics Education Society joined with another group in what Darwin later described as "a shadowy connection," and it raised the troublesome question of how far the organisation should go in working with the much more extensive pronatalist maternal and infant welfare movement that had expanded rapidly before and during the war.[17] Darwin was clearly under considerable pressure from some in the Society to take a more interventionist role, but neither he nor his mainline constituents were oriented towards the saving of infant life, especially among the lowest classes who would be the greatest beneficiaries of social programs.

Even before the war, Darwin, faced with demands to broaden the scope of the eugenic mission in improving the race, warned that eugenics must be wary of alliances with social reformers who might limit its goal of focusing upon "inborn" characteristics. At the same time, however, he admitted there was still a great deal of uncertainty about the interaction of heredity with environment. While he clearly believed that nature was overwhelmingly predominant, there could be no heredity without environment and no unit of measurement existed that could be employed for comparative purposes. Rather than squander whatever influence they might have in persuading the social reformers who were in the ascendancy to consider heredity in formulating their collectivist programs, eugenicists "at present ... should, as far as possible, avoid such phrases as the relative influence of heredity and environment, whilst always holding in view the relative possibilities of doing good by attending to heredity and to environment."[18]

It is clear that Darwin was pulling the Eugenics Education Society in two directions during the war as he struggled to preserve its mainline Galtonian principles while leading it, for admittedly political reasons, into uncomfortable alliances with such organisations as the National Association for the Prevention of Infant Mortality, the Association of Infant Welfare Centres, the National Institute of Mothercraft and, perhaps the most distasteful of all, the working class Women's Co-operative Guild. Conceding that the relative influence of heredity and environment was, as critics charged, too imprecise to spurn entirely the social improvements advocated by such groups, Darwin warned that "we must keep harping on the inequality of men as regards their inborn qualities, and we must keep repudiating environmental reform as a practical method insuring racial progress in the future." But with his characteristic caution he added that it must be done with moderation and understanding dictated by the practical exigencies of the times.[19]

Not surprisingly, although individual eugenicists were active in a number of reform organisations, the Eugenics Education Society did little more than lend its name to their causes, and after 1918 discreetly withdrew from what had been an awkward, expedient association. Eugenic complaints about the cost and futility of social welfare services, muted during the war, were quickly revived. Calculating the "racial effects" of the war, Darwin in 1919 pleaded with social workers and other advocates of public assistance not to encourage by their largesse the output of more children from poor homes whose proliferation would further undermine the heavy task of race recovery.[20] But the class eugenics of the pre-war years were now substantially diluted by democratic political realities on the one hand and, on the other, by advances in genetics that called into question a number of basic mainline assumptions about class, family and heredity.

For Darwin, whose understanding of genetics was fairly limited, there was still no doubt that inherited ability remained clustered primarily among the more successful, mainly middle classes, even if it perhaps extended downward to a much wider range of the population than previously believed. Nevertheless, there was little agreement for several years after the war about the policy consequences of such a concession. As a result, eight years into the peace the Society could agree on little more than that members were free to co-operate as individuals with "the innumerable existing bodies already striving to improve human surroundings," but it was probably best if the organisation avoided taking any position on the matter.[21]

Increasingly, for eugenicists, the most important of these bodies were tied to the birth control movement with its expanding network of voluntary clinics, the first started in 1921, largely for eugenic reasons, by the indefatigable Marie Stopes, a life fellow of the Society, in large part for eugenic reasons.[22] Darwin recognised that he was increasingly in the minority when it came to assessing the eugenic value of birth control which he doubted would ever be adopted voluntarily "by the inferior types of the community to nearly the same extent as with the superior types." Nevertheless, it was clear to him, as to many others in the early 1920s, that until the birth control question was resolved, "eugenics could not expect to advance on a wide front."[23] As a result, between 1923 and 1926 the Eugenics Society established relationships with the voluntary birth control clinics, appointed a subcommittee to study contraceptive practices among the working classes in London, and invited medical societies to recommend what could be done for "social and racial" reasons to disseminate contraceptive information.[24] The *Eugenics Review* quickly became a leading forum for the discussion of birth control questions.

At the annual meeting in 1925, Darwin, though still harbouring doubts that, "the inefficient, the careless, the weak and the stupid" would limit their families, announced that it was probably time to endorse a more definite program than had been possible a few years earlier. The following year, after intensive discussion, the Eugenics Society's Council agreed to promote birth control as a eugenic agent to be directed primarily at the poor. To take such a step, Darwin knew, would drive off some long-time supporters, but their places, he countered, "would probably be taken by others who believed that birth control was now the most important social agency for race culture."[25]

One of these others was C P Blacker, the Paris-born, Eton-educated son of a cosmopolitan family with distinguished Peruvian and American connections. Unlike Major Darwin Blacker was a scientist who had studied zoology with Julian Huxley at Balliol before moving on to medicine and psychiatry. He recognised that the determining factors in human heredity were far more complex than the first generation of eugenicists appreciated, and in the first of his many books and pamphlets on eugenics, *Birth Control and the State* (1926), he argued that for political as well as scientific reasons, biological explanations needed to be considered in a broader social context. Blacker advocated much closer co-operation with geneticists, demographers, and birth control advocates, reflecting the interests of a group of reform eugenicists, who, while still "freighted with class-dependent biases" of their own, recognised how little was really known about the role heredity played in achievement so long as environmental conditions were so unequal. Many of them, including Blacker, who described himself as a socialist, were on the political left and committed to social reform.

At the same time however, like Blacker, they continued to believe in the social correlation of good genes and bad genes, but were persuaded that since valuable characteristics were to be found in most social groups, diversity and variation were advantages, not disadvantages. They therefore felt compelled by genetic evidence to turn eugenics from the dubious identification of innate ability primarily with one class, and to concern themselves instead with populations and the biological qualities of individuals within them.[26] For a number of reform eugenicists, the Eugenics Society was a problematic vehicle for advancing this cause.

Blacker was certain that the recruitment of reform-minded doctors, scientists, and social scientists into the eugenic ranks and persuading the Eugenics Society to support valid research was the key to transforming the organisation into a credible instrument for the serious study of the interaction of genetics and the social environment. The dropping of "education" from the organisation's name at the end of 1926 was in part related to the Society's incorporation as a non-profit company, but it troubled Darwin and other mainliners who also saw it as a symbolic retreat from the emphasis upon propaganda that had characterised the role of the Eugenics Education Society since its founding.[27] In many ways Darwin never entirely accepted the change and repeatedly complained to Blacker that in becoming more scientific and research oriented, the Society was moving too far afield from Galton's vision of preaching eugenics as a new religion of biological salvation.

In 1927, building on the Society's uneasy decision to include birth control for the poor as a eugenic objective, Blacker, with the help of his friend and reform-minded ally Julian Huxley, persuaded Darwin and the council to provide £200 to help establish a Birth Control Investigation Committee (BCIC) comprised of doctors and scientists, several of whom were not eugenicists, to study the physical, mental, and racial effects of contraception on society. The decision was a controversial but critical first step in the new direction that Blacker wished to go. It not only demonstrated the growing influence of reform eugenicists, but also provided an opening for recruiting respectable physicians to the Eugenics Society where they became an influential presence in the next decade.[28] Through the BCIC the Eugenics Society was quickly drawn into a working alliance with birth control clinics and the gathering of data evaluating existing methods of contraception.

Blacker had become interested in birth control for his poor clinic patients while still a medical student at Guy's Hospital. He was appalled by the indifference of the faculty, the lack of instruction in the curriculum, and the failure of his timid profession to take the lead in the study and regulation of contraceptives, a vital area of preventive medicine, he contended, that could help avert "a biological crisis unprecedented in the history of life."[29] For Blacker, the biological crisis he saw brewing in the differential birth rate was far greater than the crisis the nation had endured in the Great War that had cost him a younger brother and profoundly changed his life.

Although doctors did not gain the monopoly over the dispensing of birth control information that he advocated,

Blacker, who served as the BCIC's secretary until it closed down in 1939, saw the committee as a way for the medical profession to at least take the lead in the study and regulation of a practice of enormous eugenic importance.[30] Major Darwin, with his usual ambivalence, lent his cautious support to the work of the BCIC so long as it remained purely an investigative body uninvolved in birth control propaganda. At the same time he prevailed upon Sir Bernard Mallet to succeed him as president in 1928, knowing that Mallet was also wary of the Eugenics Society's growing involvement with the "fanatical birth controllers."[31]

Despite Mallet's caution, by the time of his death in 1932 the Eugenics Society was more closely tied to the birth control movement than ever and the foundations for an even more abrupt turn towards reform eugenics had been laid. In 1930 the Society had received a £70,000 bequest from a retired Australian sheep farmer, Henry Twitchin, a "queer being," as Darwin described him, who had been supporting the Society from his villa in Nice throughout the previous decade.[32] Suddenly the struggling Eugenics Society, which had survived in part from periodic cash transfusions not only from Twitchin but from Darwin and other benefactors, found itself a fairly prosperous foundation under the leadership of a focused, strong-minded new general secretary who knew how he wanted to use the Twitchin legacy.

It was obvious to Blacker that eugenic considerations in the passage of reform legislation or in the formulation of population policies would never be taken seriously so long as eugenicists were perceived as motivated by selfish class prejudices and contempt for the poor. He therefore urged Society representatives to play down the class question at all costs and to emphasise that "social status is a very inadequate index of eugenic merit ... the eugenic worth of the artisan, and *probably* [italics added] also of the working classes, is just as great as that of the professional classes and leisured rich."[33]

Nevertheless, time and again he was embarrassed by eugenicists' references to "dregs and scum" and felt compelled to warn even his more progressive allies to avoid using the term "eugenically inferior" as if it were synonymous with "lower classes" or "uneducated classes."[34]

All of this smacked of a dangerous accommodation with Labour and socialism guaranteed to horrify the still vocal advocates of mainline eugenics whom Blacker did not hesitate to describe as retrograde menaces to the cause. They were particularly offended by Huxley's 1936 Galton Lecture in which he argued that the problem for reform eugenics was the old one of trying to sort out the relative contributions of nature and nurture, which could only be accomplished through the equalisation of social conditions.[35] Darwin, who hardly fell into the retrograde class, nevertheless complained to Blacker that in stressing the need for environmental reform Huxley had lost sight of the core of eugenics, "the inborn qualities of future generations." Though Blacker personally doubted environmental improvements would enhance the innate capacity of the race very much, he defended Huxley, conceding that until some equitable base for all classes was established, the difficulties of proving the claims of heredity would remain formidable.[36]

Unfortunately, as Blacker knew, the scientific credibility of eugenics was as vulnerable to attack as its social agenda. Some of its most effective critics were biologists who were able to demonstrate that when it came to the central question of the relative contributions of nature and nurture, eugenic explanations of the former precluded having much confidence in eugenic assessments of the latter. The pre-war presumption of mainline eugenic doctrine that "like produced like" through the transmission of single Mendelian unit characters had been battered by advances in genetics. To quote Kevles, "what counted in breeding was the genes of the organism – the genotype, not the expression of them – the phenotype. One could not expect to produce superior progeny simply by breeding together phenotypically superior parents," an idea that Major Darwin and his mainline colleagues never entirely abandoned. Inheritance, geneticists recognised, was polygenic, the product of the interaction of multiple genes in ways that were only beginning to be understood. Reform eugenicists, whose political and social agendas ran the spectrum from conservative to communist, by definition rejected the scientific underpinnings of mainline doctrine, and insisted that any new eugenics had to be consistent with what was known about the laws of heredity.[37]

Those laws, moderate reformers like Blacker argued, could not be isolated from the social and physical environment in which they operated. Even severe critics like the social biologist Lancelot Hogben, he wrote, had benefited eugenics by making "many of us ... more critical than we were of some traditional postulates as to the importance of heredity, and more diffident in generalising about the genetic effects of certain eugenic measures."[38] While a minority of non-scientific mainliners, several of whom angrily resigned from the Society in the 1930s, remained unpersuaded by reform eugenics and the new directions Blacker was trying to take the organisation, others, like the wealthy investment counsellor Clinton Chance, more accurately reflected the process of reassessment and redirection that was underway before the Second World War. In a troubled letter to Blacker, Chance recalled how simple and seductive the idea of hereditary quality and selective breeding once seemed. But the closer one gets to human problems, he continued, the more elusive and complex the issue becomes. While the extremes of "good and bad" traits are usually clear, when we try to separate them from environmental and hereditary factors we face a question of scientific analysis we are not really able to answer. This is the "joint in our armour as fighting eugenicists" which leaves us vulnerable to attack, Chance complained, and "it is the inability to give satisfying answers about what to do to improve the quality of future generations that makes me soft-pedal propaganda and always advocate financial support to research of ... quality."[39]

As Blacker knew all too well, not all of his members were so welcoming of his efforts to turn the well-endowed Eugenics Society into a research foundation. Shortly before Blacker became general secretary, Major Darwin, in the Galton Lecture of 1930, reminded his audience why the founders of the organisation had included "education" in its name, and argued that it should not undertake research. There were already plenty of people engaged in that enterprise, but only one organisation to explain eugenics to the public.[40] Not only was Darwin afraid that the propagandist goals of the Society would be thwarted by the demands of scientists, but he was worried that they would divert the Twitchin bequest to research projects that had only a limited relationship to eugenics.

The Society's new-found wealth proved to be a mixed blessing as it provoked fierce disagreements about how it should be targeted to advance the eugenic cause. Darwin was very suspicious throughout the 1930s that the various groups and individuals cosying up to the Society with research proposals were only after its money and had no real commitment to eugenics. With mainline propagandists and reform-minded scientists at war with each other on the council, Major Darwin wondered if his beleaguered successor Mallet was "not beginning to curse the Society and all its works!"

By way of compromise, Darwin embraced a short-lived proposal that the Eugenics Society be divided into two bodies. The first would be an endowed, independent "Population Society" engaged in purely scientific research targeted for the scientific journals. The second, dedicated to education and propaganda, would retain the name of the Eugenics Society and would have exclusive access to the Twitchin bequest, which was left by its donor for educational purposes.[41] Under

Blacker's guidance (which frequently required all of his skills as a psychiatrist) the two functions remained connected within the Eugenics Society, but the first rapidly came to dominate, though not entirely replace, the second.

In the course of the 1930s, the Eugenics Society funded not only the Birth Control Investigation Committee and the Galton Laboratory, but the British Social Hygiene Council, the Marriage Guidance Council, the Joint Committee on Voluntary Sterilization, the Society for the Promotion of Birth Control Clinics, the National Birth Control Association and its successor in 1939, the Family Planning Association. The latter two were also housed, sometimes rent-free, in the Society's premises at 69 Eccleston Square. The same was true for the Population Investigation Committee, founded at Blacker's insistence in 1936 as an independent agency and subsidised primarily by the Eugenics Society until its move to the London School of Economics after the war.[42] Similarly, at Blacker's urging the Eugenics Society also provided small grants to the Political and Economic Planning group (PEP) established in 1931 to plan for the recovery of the economy. Eugenics Society members were very active in both groups, but Blacker deliberately played down the connection to prevent critics of eugenics from withholding their assistance.

Blacker thought that the mixed reception within the Eugenics Society to his tactics reflected deep divisions over the threat of racial degeneration. The more alarmist group, he wrote to the Oxford zoologist John Baker, tended to overreact, convinced that we, as a race, are rapidly tobogganing downhill into decrepitude and extinction."[43] A second group, in which he placed himself, held that eugenic principles are important and should be taken into more account in sociology and legislation, but did not "necessarily think the race will become extinct in a short period unless the principles of eugenics are put into practice." He thought that the Society's council was increasingly inclined to share his views and to recognise that there was still plenty of scope for eugenics even if "we do not believe, as an article of faith, in our present racial decay."[44]

Each new undertaking brought complaints. Major Darwin, as usual, went back and forth. At times he agreed with Blacker that "joint action ... must beneficially increase the possible scope of our influence," even while worrying that the "general propaganda work of the Society" was being slighted.[45] Other times he groused about the new directions, and found it difficult to figure out what they had to do with eugenics. Even the Galton Lectures seemed increasingly remote. He was especially offended by John Maynard Keynes's paper in 1937, which, in Darwin's judgement, had no bearing whatsoever on eugenics. "I thank heaven I am still a eugenist ..." Darwin wrote, even if Keynes obviously was not.[46]

Though occasionally exasperated by Darwin, Blacker seems to have replied to virtually every multi-page memorandum, article or letter of complaint that the post delivered regularly from Sussex. He explained repeatedly that it would be a long time before the Eugenics Society could really push for "the fertility of desirable stocks", as Darwin wanted.[47] In the meantime, the only hope for creating the "eugenic or racial conscience" necessary to fulfil Galton's dream of "improving the inborn qualities of future generations" was by supporting responsible scientific research that might have eugenic applicability, and by affiliating with and financing neutral demographic research organisations such as the PIC and PEP where eugenic interests were well represented. Despite his scepticism, Darwin sent the PIC £50 in 1938, more as an expression of confidence in Blacker than in the alliance system the general secretary was trying to fashion.[48]

Nothing, however, defined the transition from the mainline eugenics of the pre-World War I years to the reform eugenics of the decade prior to World War II more clearly than the Society's efforts to forge a close working alliance with the birth control movement. By discreetly supporting agencies and organisations concerned with aspects of population change and social reform, Blacker calculated that the influence of the embattled Eugenics Society would greatly exceed its small membership and help it weather the danger of being irreversibly undermined by democracy, socialism, left-wing science, and, even more ominously, Nazi "race hygiene."

Shortly after the establishment of the Birth Control Investigation Committee in 1927, Blacker had persuaded the group to provide a small grant to his friend and former classmate at Oxford, the zoologist John Baker, to undertake research on a cheap, powerful, but harmless chemical contraceptive. The Eugenics Society, enticed by the prospect of a contraceptive magic bullet so simple that it could be used reliably by even the most incompetent of couples, funnelled funds through the BCIC and began a decade-long, often contentious investment in laboratory research. This undertaking led by the end of the decade to Volpar, for voluntary parenthood, no eugenic panacea, but an effective, spermicidal vaginal gel recommended by the clinics of the National Birth Control Association and then the Family Planning Association.[49]

Blacker was frequently engaged in "unspeakable dog fight[s]" with mainline eugenicists who were opposed not only to the Society's support of Baker's work, but also to Solly Zuckerman's primate studies of the female menstrual cycle to determine the 'safe period', as well as the work of other investigators into the possibility of hormonal contraception.[50] Though endocrinology was attracting some of the most innovative reproductive biologists in the 1930s, the practical application of their studies seemed too far off to be of interest to impatient eugenicists eager for a weapon as certain as sterilisation, but more practical and acceptable to the values of the day. As one of Blacker's most persuasive allies, Sir Humphrey Rolleston, Physician-in-Ordinary to George V and

Regius Professor of Physic, argued in a memorandum to the Eugenics Society Council, a simple, soluble contraceptive such as Baker was trying to invent would, "as an achievement of negative eugenics ... have racial consequences thousands of times more important than the legalising of voluntary surgical sterilisation."[51]

Each move in the direction of funding pure research stirred up resentment among mainline eugenicists of the growing influence of scientists on the Society's council, and provoked demands that the organisation return to its roots as a propaganda agency. By the mid-1930s, however, Blacker's position was strong enough to permit him to engineer radical changes in the Society's by-laws and in the make-up of the council, so that his expanding agenda of reform eugenics could go forward relatively unimpeded. Through it all, Major Darwin remained unconvinced and repeatedly warned Blacker about investing the Society's legacy in schemes of questionable eugenic return. At the same time, however, he continued to support the general secretary, encouraged Baker in his experiments, and even sent periodic contributions to him to supplement the Society's modest grants.

While Blacker, emphasising his arguments with threats of resignation, pushed and prodded the Eugenics Society in these controversial directions, he was at the same time increasingly fearful that the very term "eugenics" was so burdened by liabilities that it might prove impossible to get a fair and objective hearing in the scientific and political worlds. As if the eugenics cause did not have enough critics, the rise of Nazism threatened to add legions more, and shortly after Hitler came to power in 1933 Blacker set about trying to distance the British movement from the odious, perverted racial policies of the Third Reich.[52]

The deep sense of foreboding that Blacker had about the likely impact of Nazi programs on the fragile credibility and acceptability of eugenics was an important factor in his efforts in 1935 to change the name of the Eugenics Society to "The Institute for Family Relations." It was a way to transform the Society into a rich and influential private foundation able to play a much more central and public role in advancing both positive and negative formulations of population policy and family planning. Like the BCIC, PIC and PEP, it was also a way to provide greater cover for the increasingly unpopular image that eugenics conjured up in the minds of critics. Not only did the term evoke new Nazi-inspired images of racial tyranny, but to socialists it meant class prejudice and bigotry; to Catholics (and Blacker was one himself) false and pernicious doctrine; and to many others, including influential scientists on the left, a joke.

Despite the demographer A M Carr-Saunders' endorsement of the idea in his 1935 Galton Lecture, many of the Society's officers were, as Blacker predicted, strongly opposed. Not only were they still defiantly protective of the eugenic appellation, but they feared such a change would of necessity lead to the support of more social welfare schemes, family allowance programs, additional maternal and child care facilities, and other plans likely to encourage the fertility of the working classes.[53]

Not to be dissuaded, Blacker, the following year, entered into merger discussions with Margaret Pyke of the National Birth Control Association as a way of saving that foundering organisation and building the framework of a more positive, family-oriented institute upon its extensive network of branches and clinics around the country. He was able to persuade a majority of the Eugenics Society's officers, many of whom also served on the governing board of the NBCA, that consolidation would offer their small, London-based enterprise "an unrivalled opportunity" to steer population measures "in a biologically desirable direction" on a nation-wide basis.

Although negotiations dragged on for nearly two years, neither Blacker nor Pyke could overcome the strong resistance within their respective organisations. While Major Darwin endorsed greater co-operation between the Eugenics Society and the NBCA, he candidly told Blacker that he was opposed to any "fusion" of the two. He was fearful that the Eugenics Society would be "snowed under" and lose its identity while the NBCA would take the Society's money without accepting its racial eugenic goal of encouraging more children "in the right places." Even the new term "planned parenthood" that Blacker was touting as more representative of the positive, more inclusive direction that the merged venture would follow, failed to alleviate Darwin's suspicions. He still believed that birth control was fundamentally dysgenic even though the fertility of the lower classes was in some cases now falling more rapidly than that of their social betters. The elements of family planning such as marital counselling, sterility problems, sex education, child guidance, and women's welfare would prove costly, he warned, and divert the Society from its primary goal of advancing the "*inborn* qualities of future generations."[54]

Blacker tried to assure Darwin and other critics that the NBCA would be "subordinate" to and "almost completely controlled" by the Eugenics Society council which would house the birth control group on its premises in Eccleston Square.[55] But even some of the general secretary's supporters engaged in the negotiations came to the conclusion that the NBCA clinics and branches were highly unlikely to accept eugenic guidance and promote eugenic policies, which proved to be exactly the case. When the leadership of the NBCA, eager for the financial contributions of the Eugenics Society, proposed introducing a resolution at their annual meeting advocating closer ties, they ran into strong resistance from the rank and file, many of them supporters of Labour, and had to fight off demands that rather than merge, the birth control organisation should sever all formal links to the Eugenics Society. Though Pyke and her colleagues held out the hope of a resolution in the future, they made it clear to Blacker that until there was a drastic change in the stated aims and objectives of the Eugenics Society as well as a change of name, nothing could be done.[56]

No one understood this better than Blacker himself who had tried to change the name of his organisation two years earlier as a first step towards merger. He was content to continue to provide annual grants to the NBCA and its clinical affiliates, and to offer at low rent the second floor of the Eugenics Society's building. In one of his last acts before joining the forces in 1939, Blacker arranged for the NBCA to stay at Eccleston Square for the duration of the war rent-free. But by then it was the NBCA that had changed its aims and objectives as well as its name, to the Family Planning Association, reflecting its expansive, more positive interests in problems of fertility and family life.

Also by then the persistent conflict between mainline and reform eugenics was essentially a dead issue. Although representatives of the old guard would rise up periodically and embarrass Blacker and his reform allies, their numbers, depleted by age and resignations, had largely been supplanted by more progressive members. For the most part, these shared Blacker's more cautious, even casuistic tactics of advancing eugenic considerations through the infiltration and financial support of more credible, less controversial groups and organisations. To be sure, at Major Darwin's urging the "eugenic flag" was kept flying during the war and at Blacker's insistence the *Eugenics Review* continued to publish. But with people like Richard Titmuss at the helm until Blacker, who had been evacuated at Dunkirk and later decorated for heroism again, returned in 1943, there was little chance that the course set in the 1930s would be reversed.

Darwin's death in 1943 was also a symbolic fading of the mainline pioneer tradition that had dominated the Eugenics

Society in its first two decades.[57] In what appears to be his last letter to Blacker, Darwin emphasised that while he had supported his friend's pragmatic alliances, he had never been comfortable with them - especially birth control. He had always much preferred the eugenics of Galton with its main goal of encouraging the fitter stock to breed more. Blacker acknowledged these differences in the obituary he wrote for Darwin in the *Eugenics Review*, noting that the Society was now involved in many questions that the former president, who had guided the organisation in its critical, early years, preferred to leave alone. But in summarising Major Darwin's contributions, Blacker also offered a succinct summary of what reform eugenics was all about:

> If the infant Eugenics Society had not emphasised the role of heredity as a determinant of phenotypical characters, the prevailing assumption that the differences between men reflected merely the differences in the circumstances of their lives might have continued to hold the field. Today, without risk of misunderstanding, we can stress the biologically selective influence of factors in our social and economic life; we can contemplate the problems of nature and nurture in explicit terms, not as antithetical factors but as variables within conditions that can be defined with increasing precision.[58]

After the war, Blacker was convinced that these reform objectives had been thwarted not by the scientific challenges of new breakthroughs in the study of human genetics, but by reactions to the grotesque "Nietschean eugenics" of the Nazis that had led to the Holocaust and discredited the eugenics movement for the foreseeable future. In other words, his worst fears about the impact of Nazism on the acceptability of the eugenic ideal had come true, and by the time he stepped down in 1952 as general secretary to go to work as the first chairman of the Simon Population Trust, he knew all too well that not even the reform eugenics that he had fashioned was politically palatable.[59]

In the inhospitable years after the war, that reform tradition continued, however, as the Eugenics Society moved in directions that would eventually lead to the establishment of the Galton Institute. In 1961 the Society issued a revised statement of its aims, stressing equally the study of heredity and environment, and a decade later the Eugenics Review gave way to the more scholarly Journal of Biosocial Science. Some reformers, like John Baker, sounding more and more like the late Major Darwin, complained that the Eugenics Society had lost sight of its goal in its desire to find acceptance in the post-war era. Though Baker found the new directions "so feeble as scarcely to be eugenic," they were, as Blacker knew, a reflection of the political, intellectual, and scientific realities that the Society had to contend with after the war.[60] But even he suspected that things had gone too far in the years after his resignation. Nevertheless, the transformation of the Eugenics Society into the Galton Institute in the four decades after Blacker's departure was, in the final analysis, the logical, pragmatic outcome of the course he had set during the twenty-one years he led the organisation.

References:

[1] Daniel Kevles, *In the Name of Eugenics*. Genetics and the Uses of Human Heredity. New York, 1985, pp. 88, 176.

[2] Leonard Darwin, "The Eugenics Movement in Great Britain" p.7 unpublished typescript in the *C. P. Blacker Archive*, Contemporary Medical Archives Centre, The Wellcome Institute for the History of Medicine, 1/5.

[3] Darwin, "The Eugenics Movement in Great Britain" p. 11. See Darwin to R A Fisher, 29 March 1932 in J. H. Bennett, ed. *Natural Selection, Heredity and Eugenics*. Oxford, 1983, pp. 12, 152; John Bowlby, *Charles Darwin. A New Life*. London, 1991, pp.447-48.

[4] Cora Hodson to C J Bond, 11 Jan 1927, *Eugenics Society Papers*, Contemporary Medical Archives Centre, The Wellcome Institute for the History of Medicine. Eug/C 31.

[5] L. Darwin to Mallet, 26 Sept. 1930. *Eugenics Society Papers*, Eug/I.15; "Birth Control Investigation Committee" (1927) Pamphlet in *Eugenics Society Papers*, Eug./D.13.

[6] Darwin to Blacker, 4 October 1930, *Blacker Archive*, 1/1.37; 17 Oct. 1/1.45.

[7] L. Darwin to Mallet, 26 Sept. 1930. *Eugenics Society Papers*, Eug/I. 15; B S Bramwell to Blacker, 21 Feb. 1935, *Blacker Archive*, 13/1.19.

[8] See *Blacker Archive*, 1/3-4.

[9] Eugenics Education Society, *Fifth Annual Report 1912-13*. London, 1913, p.8; Sociological Society, *Sociological Papers*. 3 vols. London, 1905-1917, I, p.47.

[10] See Richard Soloway, Demography and Degeneration. Eugenics and the Declining Birthrate in Twentieth-Century Britain. Chapel Hill, 1990, chaps. 1-7.

[11] Galton, "Possible Improvement of the Human Breed," p.663; Darwin, "The Eugenics Movement in Great Britain" p.10.

[12] Karl Pearson, *National Life From the Standpoint of Science*. London, 1901, pp. 27-35; *Times*, 25 August 1905.

[13] Pearson to Galton, 22, 25, 29 October 1906, 2 March 1907, Francis Galton Archives, University College University College, The University of London, 293/G; 24 Jan., 29 Feb. 1908, 293/J.

[14] Darwin to Ellis, 17 Jan. 1917, *Havelock Ellis Papers*, Stirling Library, Yale University.

[15] *Eugenics Review* 9, no. 1 (April 1917), pp. 35-41.

[16] *Eugenics Review*, 7, no. 2 (July 1915), p.96 and 6, no. 4 (January 1915), pp. 288-89.

[17] Darwin to Mallet, n.d., 1932, *Eugenics Society Papers* Eug.I.3.

[18] *Eugenics Review*, 5, no. 2 (July 1913), p. 154.

[19] Darwin, "Heredity and Environment: A Warning to Eugenists," *Eugenics Review*, 8, no.2 (July 1916), 93-122.

[20] Leonard Darwin, *The Racial Effects of Public Assistance*. London, 1919, pp. 2-3.

[21] Eugenics Society, *Annual Report 1925-26*. London, 1926, p.6.

[22] Richard Soloway "The Galton Lecture: Marie Stopes, Eugenics and the English Birth Control Movement," in *Marie Stopes, Eugenics and the English Birth Control Movement.* Proceedings of a Conference Organised by the Galton Institute, London, 1996, Robert Peel, editor (London, 1997), pp. 49-76.

[23] *Eugenics Review*, 12, no. 4 (January 1921), pp. 289-90.

[24] Hodson to Stopes, 11 December 1923, *Marie Stopes Papers*, British Library, Add. Mss. 58644; *Eugenics Society Papers*, Council Minutes, 6 November 1923, 29 January 1924.

[25] *Eugenics Review*, 16, no. 2 (April 1924): 101-04; 17, no. 2 (July 1925): 141-43; "An Outline of a Practical Eugenic Policy", *Annual Report 1925-26* pp. 3-4; *Eugenics Review*, 18, no. 2 (July 1926): 95-97.

[26] C. P. Blacker, *Birth Control and the State: A Plea and a Forecast.* London, 1926, pp. 85-86; *Eugenics Society Papers*, "Meetings 1920-1929", Officers Committee, 23 February 1927; Kevles, *In the Name of Eugenics*, pp. 174-76.

[27] Eugenics Education Society, "Council Minutes". 26 May, 13 October 1926, *Eugenics Society Papers*.

[28] 'Birth Control Investigation Committee' (1927). Pamphlet in *Eugenics Society Papers*, Eug./D./13; Eugenics Society, "Meetings 1920-1929." "Council Minutes", 23 February 1927, 24 July 1929.

[29] Blacker to Stopes, 10 August 1924, *Stopes Papers*, Add.Mss. 58655; Blacker, *Birth Control and the State*, pp. 6-7.

[30] Marie Stopes, Notes of Eugenics Society Meeting 29 March 1927, *Stopes Papers*, Add. Mss. 58644.

[31] Mallet to Darwin, 21 March 1928, *Eugenics Society Papers*, Eug./C.233.

[32] Darwin to Mallet, 1 March 1930, *Eugenics Society Papers.*, Eug./I.2.

[33] Blacker to Baker, 14, 15 February 1933, Ibid. Eug./C.10.

[34] See for example Blacker to Bramwell, 22 June 1936, Ibid. Eug./C.37. "Memorandum on the Formation of An Institute of Family Relations," (1935), Ibid. Eug./C.57.

[35] Julian Huxley, "Eugenics and Society," in *Eugenics Review*, 28, no. 1 (April 1936): 11-31.

[36] *Eugenics Review*, 32, no.1 (April 1940): 5-6; Blacker to Horder, 8 June 1936, *Eugenics Society Papers*, Eug./C.171.

[37] Kevles, *In the Name of Eugenics*, pp. 144-46, 169-70.

[38] *Eugenics Review*, 30, no.4 (January 1939): 290.

[39] Chance to Blacker, 12 February 1940, *Eugenics Society Papers*, Eug./C.64.

[40] Leonard Darwin, "The Society's Coming of Age, The Growth of the Eugenic Movement." *Eugenics Review* 21, no.1 (April 1929), pp. 9-20."

[41] Bennett, *Natural Selection*, pp. 129-32; Moore to Mallet, 15 April 1930, *Eugenics Society Papers*, Eug./I.2.

[42] British Library of Political and Economic Science, *Population Investigation Committee Archives*, VIa, Blacker to Carr-Saunders, 12 June 1936; *Eugenics Review*, 31, no. 1 (April 1939): 47. See also Blacker, "Memorandum. Future Activities of this Society" [1937], *Eugenics Society Papers*, Eug/C.27; *Eugenics Review*, 35, no. 1 (April 1943): 6-7; *PIC Archives*, VIa, Glass to Carr-Saunders, 12 December 1944, 21 February, 1 March 1945; Carr-Saunders to Glass, 29 March 1945; Cadbury to Blacker, 6, 8 September 1944; Blacker to Carr-Saunders, 8 September 1944.

[43] Blacker to Baker, 15 Feb. 1933, *Eugenics Society Papers*. Eug./C.2; Blacker to Paul Espinasse, 13, Feb. 1934. Eug./C.99.

[44] Ibid., Blacker to Baker, 15 Feb. 1933, Eug./C.2; Blacker to Paul Espinasse, 13, Feb. 1934. Eug./C.99.

[45] Darwin to Blacker, 24 January 1938, *Blacker Archive*, Box 1. Folder 4.

[46] Bennett, *Natural Selection*, p.174.

[47] Blacker to Darwin, 30 July 1936, *Blacker Archive*, Box 1. Folder 3; ibid., Typescript "Equality of Opportunity and Eugenics" 1938.

[48] Blacker to Darwin, 14 Feb. 1938, *Blacker Archive*, Box 1. Folder 4.

[49] For the Eugenics Society's investment in Volpar see Soloway, "The Perfect Contraceptive: Eugenics and Birth Control Research in Britain and America in the Interwar Years," *Journal of Contemporary History*, 30 (1995), pp. 637-64.

[50] Blacker to Huxley, 5 Dec. 1934. *Eugenics Society Papers*. Eug./C.185.

[51] *Ibid.*, Blacker to Schiller, 6 December 1934. Eug./C.306. Humphrey Rolleston, "Memorandum on Grants Recommended by the General Purposes Committee," March 1934, Eug./C.295/296.

[52] *Eugenics Review*, 25, no. 2 (July 1933): 77-78; no. 3 (October 1933): 157-59, 179-81; *Lancet*, pt. 2 (5 August 1933): 297-98. Blacker, *Eugenics in Prospect and Retrospect*. London, 1945, p. 26. For examples see C. G. Campbell to Blacker, 29 January 1936, Blacker to Byron Bramwell, 7

February 1936, Eugenics Society Ps., Eug./C.37 and Blacker's correspondence with Norman A. Thompson, Eug./C.328.

[53] See "Memorandum on the Formation of An Institute of Family Relations," Ibid., Eug./C.57; Blacker to Carr-Saunders, 11 January 1935, Eug./C.57; Blacker to Huxley, 17 May 1935, Eug./C.185. Blacker to Darwin, 24 March 1937, *Blacker Archive*, Box 1/Folder 3.

[54] Darwin to Blacker, 9 Oct. 1936, 28 April 1937, *Blacker Archive*, Box 1/Folder 3; Blacker to Horder, 3 May 1937, *Eugenics Society Papers*. Eug./C.172.

[55] Blacker to Darwin, 12 Oct. 1936, 30 April 1937, *Blacker Archive*, Box 1/Folder 3.

[56] NBCA, "Minutes of the Governing Body, National Executive Committee and Sub-Committees, 1933-1937," (23 June 1937), *Family Planning Archives*, Contemporary Medical Archives Centre, The Wellcome Institute for the History of Medicine. FPA/ A5/2, 2A. See also 21 July 1937, 23 March 1938, FPA/A5/3.

[57] See Darwin to Blacker, 25 Aug. 1939; Darwin to Collyer, 19 June 1941, *Blacker Archive*, Box 1, Folder 4.

[58] Darwin to Blacker, 19 June 1941, *Blacker Archive*, Box 1, Folder 4.; *Eugenics Review*, 34, No. 4, Jan. 1943, p. 110.

[59] C. P. Blacker, *Eugenics: Galton and After*. London, 1952, pp. 142-44.

[60] Baker, "Memo to the Eugenics Society Council," 4 October 1961, *Eugenics Society Papers*. Eug./C.13

The Eugenics Society And The Development Of Demography In Britain: The International Population Union, The British Population Society And The Population Investigation Committee

Chris Langford

Introduction

The present-day International Union for the Scientific Study of Population (IUSSP) has nearly 1800 members world-wide (the 1996 Directory of Members lists 1778) and is engaged in a very full and wide-ranging programme of population-related activities including the organisation of conferences and research seminars and the publication of research findings. Before the Second World War, and indeed until 1947, this organisation was known as the International Union for the Scientific Investigation of Population Problems (IUSIPP) or sometimes as simply the International Population Union (IPU). The organisation had been founded in 1928, in Paris, following a World Population Conference in 1927, held in Geneva, which Margaret Sanger had been instrumental in bringing about. Unlike the IUSSP today, to which individuals apply directly for membership, as individuals, the pre-1947 IPU, though it had a central, elected, governing body, had a membership which essentially amounted to simply the sum total of a series of 'national committees' each with its own members (and, effectively, membership criteria). The British component of the IPU, established in 1928, was the British Population Society (BPS); occasionally this was also referred to as the British National Population Committee, the British National Committee, or the British Population Committee. (The present-day British Society for Population Studies, which was founded in 1974, has no connection whatsoever with the British Population Society.) There was, however, another aspect of the British connection with the International Population Union in the pre-war period in that, between 1931 and 1937, the two major officers of the overall international organisation, the President and the Honorary General Secretary and Treasurer, were both British and the organisation's headquarters was in London.

The Population Investigation Committee (PIC) was established in 1936 and is still in existence today (I am a member of it). Arguably the committee's most important activity now is the publication of the journal *Population Studies*; it also provides, fairly modest, grants to individuals and organisations for a variety of population-related purposes. However, during the late 1930s and through the 1940s and 1950s, the PIC was probably Britain's foremost demographic research organisation.

The main object here is to trace the development of demography in Britain - or at least certain aspects of this development - especially in the 1930s and 1940s, through the activities of the IPU, the BPS and the PIC, and to consider the possible influence of the Eugenics Society in this development. According to that Society's first President, Sir Francis Galton, 'Eugenics is the science which deals with all influences that improve the inborn qualities of a race; also with those that develop them to the utmost advantage' (see Blacker, 1952, p. 17). In the case of the PIC there was, at least at the outset, a direct link with the Eugenics Society in that the latter had actually established the PIC in the first place; however there was no such formal connection where the IPU and the BPS were concerned. Two kinds of evidence of the possible influence of the Eugenics Society on other bodies will be considered: the extent to which their intellectual concerns reflected those of the Eugenics Society and/or the extent to which they shared members with that organisation. Throughout, for convenience, both the original Eugenics Education Society, founded in 1907, and its renamed successor, the Eugenics Society, will be referred to simply as the Eugenics Society.

Strictly, the first part of the title of this paper ('The Eugenics Society and the development of demography in Britain') implies the need to discuss more than just the three bodies referred to in the second part of the title. One ought to consider, in addition, at the very least, the possible influence of the Eugenics Society on the work of Britain's official statisticians, as well as, of course, the considerable body of demographic work and debate that occurred within the framework of the Eugenics Society itself, through, for example, meetings and through the materials published in *The Eugenics Review*. Richard Soloway (1990) and Simon Szreter (1996) have however already written a good deal about these other aspects of the matter. Here, therefore, I shall confine myself to an account of the IPU, the BPS and the PIC.

The foundation and activities of the IPU and the BPS and the extent and nature of British and Eugenics Society involvement

In 1927 a World Population Conference was held in Geneva. Margaret Sanger, the American birth control activist, was instrumental in bringing the conference about and she subsequently edited the report on the conference (Sanger, 1927). The conference President was Sir Bernard Mallet, a former Registrar General of England and Wales, and there were 30 other British participants. The list included so many famous names, or at the very least names which will appear again in this account, it is probably worth reproducing in full. The other British participants in the conference were: C P Blacker, Lord Buckmaster, Mabel Buer, A M Carr-Saunders, Sir Charles Close, Harold Cox, F A E Crew, Lord Dawson of Penn, C V Drysdale, Binnie Dunlop, Havelock Ellis, R A Fisher, Morris Ginsberg, J W Gregory, J B S Haldane, David Heron, Cora Hodson, Sir Thomas Horder, Julian Huxley, The Very Reverend Dean Inge, J Maynard Keynes, E J Lidbetter, F H A Marshall, F J McCann, G H L F Pitt-Rivers, Gladys Pott, Sir Humphrey Rolleston, Percy Roxby, H Sutherland and H G Wells. One is almost tempted to add to this list the name of a German participant in the conference, R R Kuczynski, in that Kuczynski was later to be the first ever appointee to a teaching post in demography at a British university (and a member of the PIC).

At the 1927 conference it was decided that a permanent international organisation devoted to the scientific study of population should be set up and a committee was established, under the chairmanship of Raymond Pearl (of The Johns Hopkins University, Baltimore), to implement this decision; the British members of the committee were F A E Crew, who was Secretary, and Sir Bernard Mallet. The International Union for the Scientific Investigation of Population Problems, or International Population Union, was formally constituted in Paris in July 1928. Pearl was elected first President; Sir Bernard Mallet was one of the three Vice-Presidents, and Honorary Treasurer. The union was made up, not of individual members, but of a number of National Committees. Despite some press reports, even before the 1927 conference began, suggesting that a neo-Malthusian organisation was envisaged (understandable, perhaps, given Margaret Sanger's involvement) and indeed much the same suspicion on the part of some conference participants, who objected to this either because of a wish to preserve scientific neutrality or simply because they themselves held anti-neo-Malthusian views

DEVELOPMENT OF DEMOGRAPHY

(Corrado Gini of Italy, for example), the whole formal ethos of the new organisation was determinedly a strictly scientific one (IPU, 1932). The statutes of the organisation, which were published in the first issue of the *Bulletin of the IUSIPP* in October 1929, made clear that the union confined itself 'solely to scientific investigation in the strict sense' and refused 'either to enter upon religious, moral, or political discussion, or ... to support a policy regarding population, of any sort whatever, particularly in the direction either of increased or of diminished birth rates'. It has been suggested, not surprisingly, that Margaret Sanger was rather disappointed with the way the new organisation developed (IUSSP, 1985, p.5).

Either at Paris itself, or very soon afterwards, three Research Commissions were established by the IPU. The first was concerned with 'population and food supply', the second with 'differential fertility, fecundity, and sterility' and the third with 'vital statistics of primitive races'. The British (or, at least, British-based) members of these Research Commissions were: first commission, Sir Henry Rew; second commission, A M Carr-Saunders, F A E Crew (chairman), Eldon Moore (secretary), and T H C Stevenson; and, third commission, B Malinowski (vice-chairman) and G H L F Pitt-Rivers.

The British component of the IPU, the British Population Society, was established in October 1928 following the publication of a letter in *The Times* in September from Sir Bernard Mallet, which was then reprinted in the October 1928 issue of *The Eugenics Review*. The temporary address of the British National Committee was given in *The Eugenics Review* as c/o the Eugenics Society, who published that journal, and it was stated that Eldon Moore, its editor, would act as Honorary Secretary of the new organisation. Also published in *The Eugenics Review* were the names of individuals 'among those who are being invited to become members'. A similar list of names appeared subsequently as the actual membership of the British Population Society in the *Bulletin of the International*

Union for the Scientific Investigation of Population Problems (Volume 1, No. 3, January 1930). This gave the membership as: Sir Bernard Mallet (Chairman), E Moore (Honorary Secretary), Sir William Beveridge, A L Bowley, A M Carr-Saunders, Sir Charles Close, F A E Crew, R A Fisher, J W Gregory, D Heron, J S Huxley, The Very Reverend Dean Inge, Sir Arthur Keith, J M Keynes, B Malinowski, F H A Marshall, M Pease, G H L F Pitt-Rivers, Sir Humphrey Rolleston and Sir Josiah Stamp.

The second 'general assembly' of the IPU (the first had been at Paris) and its first conference were held in London in June 1931. Twenty-three papers were presented and there were reports from two of the Research Commissions and a number of National Committees. The scope of the conference was wide, with papers on medical, biological, anthropological and agricultural aspects of population, as well as on social and technical demography: not only A J Lotka on 'The structure of a growing population' and F W Notestein on 'The relation of social status to the fertility of native-born married women in the United States' but F A E Crew on 'Some experiments on populations of mice', J Warming on 'Trends in agricultural production in Denmark' and T Kemp on 'The significance of blood-grouping in anthropology'. The report on the conference, edited by G H L F Pitt-Rivers, was published the year after the conference as *Problems of Population* (Pitt-Rivers, 1932).

At the time of the 1931 conference, the IPU statutes were modified, and new officers were elected. Raymond Pearl had indicated his intention of resigning the presidency some time before. Sir Charles Close became President and G H L F Pitt-Rivers, Honorary General Secretary and Treasurer. (At some stage, and certainly from late 1934 onwards, E C Rhodes seems to have been added, as 'Assistant Secretary' of the IPU.) Sir Bernard Mallet remained a vice-president, now one of seven. Sir Charles Close and G H L F Pitt-Rivers (plus Rhodes)

DEVELOPMENT OF DEMOGRAPHY

continued in these posts until the IPU conference in Paris in 1937.

Accommodation for the IPU was promised at the London School of Economics (LSE) by the School's Director, Sir William Beveridge, after certain rebuilding work had been completed. In the meantime, an office was provided by the Royal Geographical Society in London. Previously, the IPU had effectively operated from Raymond Pearl's office at The Johns Hopkins University in Baltimore.

The original intended venue for the 1931 conference had been Rome; and indeed a much larger population conference was held in Rome in that year, in September; the proceedings of that conference, edited by Corrado Gini, filled ten volumes (Gini, 1933-1935). The official IPU conference was switched to London because of a dispute between the Italian National Committee (whose chairman was Corrado Gini) and the rest of the organisation. The precise nature of the dispute is unclear but it seems to have had its origins in the fact that further financial support which the IPU had expected to receive from American research councils (substantial help was received from the Milbank Memorial Fund) did not materialise. It has been suggested that one element in the situation may have been that the research councils reacted negatively to the idea of the IPU holding its first conference in Rome, given the Fascist government's, and Gini's, pronatalism, and the perception that Gini himself was too closely associated with the Italian government and its policies; however, personal and scientific disagreement between Pearl and others, within as well as outside the United States, may well also have played a part (Federici, 1984; IUSSP, 1985). In any event, the Italians boycotted the London meeting and thereafter viewed the decisions taken there as illegal.

Starting in October 1929, a *Bulletin of the International Union for the Scientific Investigation of Population Problems* had been published 'for the information of the members of the

International Union', from Raymond Pearl's office at Johns Hopkins. There were ten issues altogether, with the last in July 1931. Subsequently, after the IPU's move to London, a fully-fledged journal was established, under the title *Population*, the first issue of which appeared in June 1933. The journal was published in London by George Allen and Unwin. There were nine issues altogether, the last in January 1939. The editor throughout was E C Rhodes, Reader in Statistics at the London School of Economics. Forty-nine articles appeared altogether, four in French, three in German, and the rest in English. The largest number of papers came from British contributors but there were several papers also by American and by Dutch authors, as well as other contributions.

The list of members of the BPS given in the report on the 1931 London conference was essentially the same as the one which has already been provided. There were three new members, however: Mabel Buer, C B Fawcett and Gladys Pott; and F H A Marshall was omitted from the 1931 list. It was also reported in the conference proceedings that J W Gregory had died a year after the conference. Thus the situation then was not very different from that in 1930 when Sir Bernard Mallet had reported in the *Bulletin of the IUSIPP* (Volume 2, No. 1, September 1930) that 'Appeals in the press failed to bring in new adherents and the Society is now confined to the 20 or 30 [nearer the former, I think] eminent scientists and statisticians, whose names have been published. ... [M]ore activity must be shown in securing new members and in securing funds ...'. There presumably must have been such a recruitment drive since a BPS pamphlet dated May 1937, giving the rules, objects, constitution, etc. of the society, listed 40 members (as well as the London School of Economics as an institutional member) and 26 'associate members' (the latter had no voting rights and were not regarded as members of the IPU). From that same pamphlet, as well as issues of the journal *Population*, and conference proceedings, letterheads, etc., we also know that, in

DEVELOPMENT OF DEMOGRAPHY 89

addition to Sir Bernard Mallet, Sir Charles Close (later, Arden-Close), C B Fawcett and G H L F Pitt-Rivers were all Chairman of the BPS at some stage; and that, in addition to Eldon Moore, Mabel Buer and C Conyers Morrell acted as Honorary Secretary. We also have the names of a number of individuals who were, non-office-holding, members of the Executive Committee of the BPS at some point, viz. C P Blacker, A L Bowley, Mabel (M C) Buer, A M Carr-Saunders, Sir Charles Close, C B Fawcett, R A Fisher, R Ruggles Gates, D V Glass, C B S Hodson, G H L F Pitt-Rivers and E C Rhodes. None of the above lists is necessarily exhaustive, however. Table 1 shows all those individuals that I have been able to establish were members of the BPS at some stage and whether they were BPS office holders or members of the Executive Committee.

The second IPU conference was held in Berlin in 1935; it had originally been planned for 1934 but was postponed. There were a dozen British participants including Sir Charles Close and G.H.L.F. Pitt-Rivers, still, respectively, President and General Secretary of the IPU. David Glass, who was also present, subsequently wrote a highly critical account of the conference, which was published in *The Eugenics Review*, indicting both the scientific content and the racialist atmosphere that prevailed (Glass, 1935). He also reported that many scholars, including virtually all American demographers, had boycotted the conference. The proceedings of the conference, edited by Harmsen and Lohse, were published in 1936 (Harmsen and Lohse, 1936).

The IPU Paris conference of 1937 was much less controversial, though by no means without its troubles. G H L F Pitt-Rivers prepared, and presumably circulated, a report for the 'general assembly' that was to take place (this was ultimately the IPU's sovereign body) in which, *inter alia*, he more or less called for the expulsion from the IPU of the National Committees of Spain, Italy, Poland and Czechoslovakia (Pitt-Rivers, 1937). Non-payment of dues was one of the reasons given; however, in the case of Spain, where the civil war was still in progress, and in the case of Czechoslovakia, where Pitt-Rivers alleged there was discrimination against the German-speaking minority, political factors were clearly involved. It does not seem that Pitt-Rivers' report was actually discussed by the 'general assembly'; it was considered, at least in a formal sense, at a meeting of the Executive Committee of the union the day before, but it is clear from the minutes of that meeting that effectively they ignored it. He later wrote (on the subject of the German-speaking population of Czechoslovakia) of '- the suppression of my reports on the subject at the International Population Congress in Paris in July, 1937' (Pitt-Rivers, 1938, p.62).

At the Paris conference, new officers of the IPU were elected: Adolphe Landry became President in place of Sir Charles Close and Georges Mauco replaced G H L F Pitt-Rivers as General Secretary. Sir Charles Close became one of the seven vice-presidents of the IPU. The proceedings of the conference were published, in eight volumes, in 1938 (IPU, 1938).

What part did the Eugenics Society play in the formation and development of the IPU and the BPS? Table 1 lists all those individuals that I have been able to establish were members of the BPS at some stage (there may have been others) and whether they were office holders in that organisation or members of its Executive Committee; the table also shows whether these individuals were present at the 1927 Geneva conference, which gave rise to the IPU and the BPS, and, so far as I have been able to determine, whether they ever belonged to the Eugenics Society, when they first joined, and whether they were ever an office holder or member of the Eugenics Society council. Table 2 lists British participants in the 1927 Geneva conference and indicates, in similar fashion, the extent of their involvement in the BPS and the Eugenics Society.

DEVELOPMENT OF DEMOGRAPHY

Table 1. Those known to have been members of the BPS at some stage and the extent of their involvement in the BPS and certain other activities[a]

Name	BPS office holder (O) or Executive Committee Member (C)? *	At 1927 conference ? (Y=yes)	Eugenics Society office holder (O), council member (C) or member (M)? *	Joined Eugenics Society no later than?
Founder Members[b]				
Sir William Beveridge			C	1928
A.L. Bowley	C			
A.M. Carr-Saunders	C	Y	O,C	1913
Sir Charles Close	O,C	Y	C	1928
F.A.E. Crew		Y	C	1924
R.A. Fisher	C	Y	O,C	1913
J.W. Gregory		Y		
David Heron		Y	C	1909
Julian S. Huxley		Y	O,C	1924
The Very Revd. W.R. Inge		Y	C	1909
Sir Arthur Keith				
J. Maynard Keynes		Y	O,C	1920
B. Malinowski				
Sir Bernard Mallet	O	Y	O,C	1919
Eldon Moore	O		M	1927
Michael Pease			C	1920
G.H.L.F. Pitt-Rivers	O,C	Y	C	1920
Sir Humphrey Rolleston		Y	O	1933
Others joining before 1930[c]				
F.H.A. Marshall		Y	M	1934
Sir Josiah Stamp			C	1928
Those joining 1930 or later				
T.T. Behrens				
C.P. Blacker	C	Y	O,C	1927[d]
C.I. (misprint for J) Bond			O,C	1909
B.S. Bramwell			O,C	1922
C.H.I. Brown				
Mabel (M.C.) Buer	O,C	Y		
A.T. Culwick				
C.B. Fawcett	O,C			
H.J. Fleure			C	1920
R. Ruggles Gates	C		O,C	1920
D.V. Glass	C		O,C	1937
G. Talbot Griffith				

Cora (C.B.S.) Hodson	C	Y	O,C	1926
Mrs How-Martyn			M	1928
W.W. Jervis				
B.N. Kaul				
J.R. Marett				
K. Mason				
Elton Mayo				
G.F. McCleary				
C. Conyers Morrell	O		M	1928
The Rev. S.T. Percival				
Gladys Pott		Y		
E.C. Rhodes	C			
S. Rowson				
J. Rumney				
J.R. Ryder				
H.W. Seton-Karr			M	1929
E.C. Snow				
Marie Stopes			M	1913
Griffith Taylor				
G.E. Whitrod				

* O = office holder on committee or council; C = non-office-holder on committee or council; O,C = both, at different times. Those holding the office of Honorary Secretary or General Secretary have been recorded as O. O and C of course imply membership; M has been used to indicate membership only. In the case of the Eugenics Society, the distinction between members and associates and, later, between fellows and members has been ignored; all have been regarded as members.

[a] This table shows BPS members. A BPS pamphlet dated May 1937 also lists 'associate members' (who were not members of the IPU), as follows: H. El-S. Azmi, A. M. Close, B. Dunlop, W. Edge, M. El-Darwish, R.E. Enthoven, C. Daryll Forde, Lindley Fraser, The Hon. Mrs. Grant, A. Hall-Hall, Miss E. Hawarden, Dorothy E. Johnston, Mrs A. Laye, A.F. Leest, The Hon. M. Lubbock, E. Lucas, D.H. McLachlan, P.E. Percival, Miriam Rothschild, C. Russell-Brown, Edith Seymour, Catherine Sharpe, E.A. Short, Eva G.R. Taylor, J.P. Williams-Freeman and M.F. Wren.

[b] According to a document in the PIC files dated August 1931, 'Constituted as original members with power to add to their numbers at a meeting held on October 26, 1928'.

[c] That is, included in the list of members published in the January 1930 issue of the *Bulletin of the IUSIPP*.

[d] A C. Blacker of Torquay was listed as a member in 1920.

Table 2. British participants in the 1927 World Population Conference and the extent of their involvement in the BPS and the Eugenics Society.

Name	BPS office holder (O), Executive Committee member (C) or member (M)?*	Eugenics Society office holder (O), council member (C) or member (M)?*	Joined Eugenics Society no later than?
C.P. Blacker	C	O,C	1927[a]
Lord Buckmaster			
Mabel Buer	O,C		
A.M. Carr-Saunders	C	O,C	1913
Sir Charles Close	O,C	C	1928
Harold Cox		M	1927
F.A.E. Crew	M	C	1924
Lord Dawson of Penn			
C.V. Drysdale		C	1909
Binnie Dunlop		M	1909
Havelock Ellis		C	1909
R.A. Fisher	C	O,C	1913
Morris Ginsberg			
J.W. Gregory	M		
J.B.S. Haldane			
David Heron	M	C	1909
Cora Hodson	C	O,C	1926
Sir Thomas Horder		O	1932
Julian Huxley	M	O,C	1924
The Very Revd. Dean Inge	M	C	1909
J. Maynard Keynes	M	O,C	1920
E.J. Lidbetter		C	1910
Sir Bernard Mallet	O	O,C	1919
F.H.A. Marshall	M	M	1934
F.J. McCann			
G.H.L.F. Pitt-Rivers	O,C	C	1920
Gladys Pott	M		
Sir Humphrey Rolleston	M	O	1933
Percy Roxby			
H. Sutherland			
H.G. Wells			

* See note on Table 1
[a] See note d on Table 1

A list of participants in the 1927 Geneva conference was provided in the conference proceedings, so one hopes that this information is correct. So far as the BPS is concerned, in connection with which most sources of information have already been detailed (but see also Table 1, notes b and c), I believe that the information provided is correct but I cannot be sure of its completeness: there could conceivably have been other members of the BPS, of whom I have not become aware, and also other office holders and Executive Committee members. In the case of the Eugenics Society information has been taken from its Annual Reports and from issues of *The Eugenics Review*; there are also for some years free-standing printed membership lists produced by the society (these latter I have seen at the Contemporary Medical Archives Centre at the Wellcome Institute for the History of Medicine, in London). These records do not provide a complete picture, however: for some years, and sometimes for several years running, there may be full lists of members, but for many years there may only be listings of officers of the Eugenics Society or of council members (and occasionally not even this); sometimes 'new members' are indicated. Thus, the information on Eugenics Society involvement presented in Tables 1 and 2 may be incomplete: someone may wrongly be shown as never a member simply because of a gap in the record (or a mistake on my part - there is a huge volume of material to go through) or they may have joined the Eugenics Society before the year shown (hence the cautious heading 'joined Eugenics Society no later than'); it is also conceivable, though less likely, that someone may have been an Eugenics Society office holder or council member but that this is not indicated. A strict view has been taken in determining the latest possible year of joining the Eugenics Society; for example, those listed as members in the first annual report of the Eugenics Society were recorded as having joined no later than 1909 though it is conceivable they joined as early as 1907.

Most of the people of interest in this account seem to have had a fairly clear-cut relationship with the Eugenics Society in the sense that they either joined the organisation or did not,

DEVELOPMENT OF DEMOGRAPHY 95

but, if they did join, they then remained members for many years, perhaps for life. However, a few did not conform to this pattern. David Heron was recorded as a member (indeed, council member) in the first annual report of the Eugenics Society but seems to have disappeared from the lists after a year or so. Harold Cox may well also have been a member for only a relatively short period (I have found him on only one list but that is not necessarily definitive). Sir William Beveridge too seems to have been a Eugenics Society member and, part of the time, council member, for a few years in the late 1920s and early 1930s, but then to be dropped from the lists. Beveridge obviously maintained some kind of relationship with the Eugenics Society subsequently, though. He was invited to deliver the society's showpiece Galton Lecture in February 1943 (his title was 'Eugenic aspects of children's allowances') in which, interestingly, he argued, somewhat apologetically, that, despite appearances, his proposed system of child allowances would bring eugenic benefits since although they would be paid to all parents of more than one child at a flat rate per child and thus might be thought dysgenic in that they would act as a greater encouragement to childbearing for the poor than the better off, the allowances would not be dysgenic in their effect as they could only influence 'parents who take some thought over the begetting of their children' (I take the argument to be that only the respectable element among the relatively poor would be influenced and that they were eugenically acceptable - the behaviour of the unrespectable poor would not be affected); Beveridge did believe that other measures to encourage more childbearing by higher status, and putatively higher quality, groups would be desirable, however (Beveridge, 1943).

It is clear from Table 2 that Eugenics Society members were heavily involved in the 1927 Geneva conference, which was to lead on to the foundation of the IPU and the BPS. The British participants included a substantial number who were already members of the Eugenics Society as well as others who would become members subsequently (almost two-thirds of British participants were past or future members of the Eugenics Society); moreover, virtually all of them were 'prominent' Eugenics Society members in the sense of being, at some stage, an Eugenics Society office holder or council member. This heavy involvement of Eugenics Society members in the Geneva conference is not especially surprising given the origins of the conference. The conference programme had been arranged by a committee, based in London, chaired by Sir Bernard Mallet, who then went on to be President of the conference as such. Sir Bernard Mallet was already then a leading member of the Eugenics Society (probably more or less its President-designate) and would soon become the society's President (in 1929, or possibly late-1928). It might be added (though this may or may not have any real significance) that Margaret Sanger had earlier joined the, British, Eugenics Society (not later than 1920) and would later become a life member.

In due course, as has been noted, Sir Bernard Mallet established the British Population Society; and again the Eugenics Society was heavily involved. *The Eugenics Review* carried information about the intended new organisation; its founding meeting in late October 1928 took place on Eugenics Society premises; its address was given as c/o the Eugenics Society; and the editor of *The Eugenics Review* acted as secretary of the new organisation. Sir Bernard Mallet himself, by that time, was already, or very soon to be, President of the Eugenics Society. It may be seen also from Table 1 that very many of the founding members of the BPS were Eugenics Society members; moreover, that virtually all of these were, at some stage, Eugenics Society office holders or council members, the only exception being Eldon Moore, who was however editor of the Eugenics Society journal. Table 1 also shows that, of the 15 individuals whom I have been able to establish were BPS office holders or members of its Executive

DEVELOPMENT OF DEMOGRAPHY 97

Committee at some point, only four, A L Bowley, Mabel Buer, C B Fawcett and E C Rhodes, seem not to have been Eugenics Society members; moreover, of the eleven who were Eugenics Society members, nine were Eugenics Society office holders or council members (and one of the other two was Eldon Moore).

So far as the IPU itself was concerned, it has already been noted that between 1931 and 1937 the principal officers of the organisation were British and its headquarters was in London. Both Sir Charles Close (the IPU President) and G H L F Pitt-Rivers (General Secretary) were prominent members of the Eugenics Society (though E C Rhodes - editor of the IPU journal and Assistant Secretary - was apparently not a member at all).

The foundation and activities of the PIC and the involvement of the Eugenics Society

The first meeting of the Population Investigation Committee was on 15 June 1936. The meeting was held at the London School of Economics where indeed most subsequent meetings have taken place though for a period during the war when the LSE was evacuated to Cambridge the PIC usually met in the rooms of the Eugenics Society in London. The PIC was set up by the Eugenics Society but was intended from the outset to be independent of it: in the words of the Eugenics Society Annual Report for 1936-37 'The Population Investigation Committee, though called together by the Society, is not a sub-committee of the Council, nor is it subordinate to the Society. It is an autonomous joint committee'.

Within the Eugenics Society an important focus of disagreement in the 1930s (and earlier) was the suggestion by some that the society had been too much concerned with 'negative' eugenics (the discouragement of childbearing by the less satisfactory) and too little with 'positive' eugenics (the encouragement of childbearing by the more satisfactory). In late 1934 some of those who took this view persuaded the society to establish the Positive Eugenics Committee: A.M. Carr-

Saunders, then Charles Booth Professor of Social Science at the University of Liverpool, was its chairman and C P Blacker, the Eugenics Society General Secretary, its secretary. The committee decided that it wished to collect evidence on the measures taken to increase fertility in a number of European countries. D V Glass was taken on as research assistant to do this work (he was the second or possibly the third person approached with the offer of the job) and the report on this research appeared as Glass' *The Struggle for Population* which was published in 1936. In some ways this committee may be seen as a forerunner of the PIC itself; certainly the three individuals mentioned were to be very important, in much the same capacities, in the establishment and development of the PIC.

On 16 February 1935 Carr-Saunders gave the Galton Lecture of the Eugenics Society: his title was 'Eugenics in the light of population trends': the lecture was published in the *Eugenics Review* in April 1935. He stressed the importance of basing policy on hard information. He urged a greater emphasis on positive rather than negative eugenics (he felt that there was a danger of the Eugenics Society becoming known as 'The Society for the Detection of Persons Undeserving of Posterity'). He drew attention to the fact that the net reproduction rate in Britain was below one and that the population would soon begin to decline unless fertility increased which he obviously did not expect to happen. Enid Charles had painted a similar picture in her *The Twilight of Parenthood* published the previous year. He felt that although so far people had not really become aware of the impending population decline they soon would and then there would be pressure for policy measures to increase fertility. The Eugenics Society should make ready for this time so that it would be in a position to recommend a policy based on appropriate eugenical principles: 'What is required is that some organisation, which has the whole population situation under review and desires to

DEVELOPMENT OF DEMOGRAPHY 99

construct an adequate programme, should examine all the proposals made to deal with these difficulties, and weave them into a coherent population policy'.

It is possible that Blacker suggested the theme of his Galton Lecture to Carr-Saunders; Blacker himself may have had the idea from Enid Charles (both of these points seem to be implied in an undated, unsigned note in David Glass' handwriting). In any case certainly after the Galton Lecture and possibly beforehand Blacker and Carr-Saunders together mapped out the follow-up to the lecture. They decided that two new organisations were needed: one to concern itself strictly with research and the other with the formulation of policy. The PIC emerged as the research body they envisaged; later, in 1938, a separate Population Policies Committee was established.

Following the Galton Lecture the Eugenics Society decided to implement Carr-Saunders' suggestions and established a committee to recommend how this might best be done. The committee consisted of the society's then President, Lord Horder, and C P Blacker, A M Carr-Saunders, Eva Hubback and J S Huxley. It was decided that a new organisation should be set up and that this should be strictly scientific in nature and independent of the Eugenics Society: only in this way could it hope to become intellectually credible and be seen as impartial; moreover these features might well enhance its prospects of attracting funds and make it easier to enlist the co-operation of other bodies. It was decided that a number of organisations should be invited to nominate representatives to the new body and that some individuals should be asked to become members in their own right. The Eugenics Society made available the sum of £250 and promised a further £250 should this be required (this was to be the first of many grants made by the Eugenics Society to the PIC). The new body, the Population Investigation Committee, duly met in June 1936. Its headquarters was in the offices of the Eugenics Society which provided secretarial assistance (this situation continued until 1946 since which time the PIC has been based at the LSE). The aims of the PIC, in the words of its first annual report, were '-to examine the trends of population in Great Britain and the Colonies and to investigate the causes of these trends, with special reference to the fall of the birth-rate'.

At the first meeting of the PIC Carr-Saunders was elected chairman of the committee and Blacker its general secretary. It was also agreed that Glass (who was not present) would be invited to become the committee's research secretary: this was to be a full-time paid appointment. There were nine people present at this first meeting; however the membership of the committee expanded in the months that followed. At the time of the first annual report of the PIC its membership was as follows: A M Carr-Saunders* (Chairman), Sir Walter Layton (Treasurer), C P Blacker* (Honorary Secretary), D V Glass * (Research Secretary), E Holland* (representing the British College of Obstetrics and Gynaecology), L S Penrose (the Medical Research Council), H D Henderson (the Royal Economic Society), S. Churchill* (the Society of Medical Officers of Health), Sir Charles Close* (the British Population Society), F Frederick (the College of Nursing), Lord Horder*, E M Hubback* and J Huxley* (all representing the Eugenics Society) and (as individuals) C Clark, L Hogben, R R Kuczynski, E M H Lloyd, D H McLachlan, T H Marshall, Lady Rhys Williams and J Young. Nine of these 21 individuals were members of the Eugenics Society in 1937 (these have been marked with an asterisk; Blacker has been counted as a member, though strictly an employee) and some others had been or would become members. Kuczynski, for whom I have found no record of Eugenics Society membership, has been described by Soloway (1990, p. 248) as an Eugenics Society 'fellow traveller'. Hogben, on the other hand, was an avowed critic of eugenics.

Carr-Saunders, Blacker and Glass were all to remain members of the PIC until they died. In 1936 Carr-Saunders was Professor of Social Science at the University of Liverpool; from 1937 to 1956 he was Director of the London School of Economics; he remained chairman of the PIC until 1958 when David Glass took over; he died in 1966. Blacker was a practising psychiatrist: at various times he was associated with the Department of Psychological Medicine at Guy's Hospital and with the Bethlem Royal and the Maudsley Hospitals. He was general secretary of the Eugenics Society from 1930 to 1952; this was, at least until the war and possibly later (I have not checked the later records), a paid half-time post. He continued as general secretary of the PIC until a few months before he died in 1975. Glass was full-time research secretary of the PIC from its foundation until the war and then again between 1944 and 1946; in 1946 he was appointed Reader in Demography at the LSE but continued as research secretary to the PIC on a part-time basis; in 1948 he was appointed Professor of Sociology at the LSE, relinquished the research secretaryship (E Grebenik took over) and became vice-chairman of the PIC; he became chairman of the PIC in 1958 and remained so until his death in 1978.

In 1936 the PIC set out to develop a 'questionnaire on fertility'. This was intended to be completed for ever married women by doctors and allied personnel and covered birth control practice as well as fertility. The questionnaire went through several drafts and was extensively tested before a final version was settled on. The project then came to an end because of the war; however, the questionnaire later provided a basis for that used in the survey carried out by Lewis-Faning in 1946-47.

In 1937 there was a debate in the House of Commons on a motion to the effect that the threatened decline of the population might constitute a danger and asking for an enquiry and report by the government. Following this debate the PIC was invited to enter into discussions with the General Register Office on how official statistics might be improved. Several meetings took place, at which the PIC was represented by Carr-Saunders, Blacker and Glass. They suggested a number of new analyses of available material but urged also that some additional information be collected at birth and death registration. Most of their suggestions were accepted and embodied in the Population (Statistics) Act 1938. Perhaps the most important new items of information were (at birth registration) age of mother at birth, birth order and mother's date of marriage.

In 1937 G Leybourne was appointed 'full-time associate research worker' by the PIC; she worked on the relationship between the costs of education and family size and went on to publish *Education and the Birth-Rate* with K White in 1940. In 1938 Glass on behalf of the PIC submitted a memorandum to and appeared in person before the Inter-Departmental Committee on Abortion; his memorandum was published as 'The effectiveness of abortion legislation in six countries' in the *Modern Law Review* in 1938. In 1938 he also published 'Changes in fertility in England and Wales, 1851-1931' and 'Marriage frequency and economic fluctuations in England and Wales, 1851-1934' as contributions to Hogben's *Political Arithmetic*. In 1938 Kuczynski, appointed that year as Reader in Demography at the LSE, began work on 'a demographic handbook of the Colonial Empire'; the PIC obtained $5750 from the Carnegie Corporation of New York in support of this project as well as a substantial grant from the Colonial Office Research Fund; the results appeared in the three volumes of Kuczynski's *Demographic Survey of the British Colonial Empire* published in 1948, 1949 and 1953 (Kuczynski died in 1947). In December 1938 the PIC published a booklet of some one hundred pages by Glass and Blacker entitled *Population and Fertility;* amongst other things this included a plea for a fertility census in which dates of birth of liveborn children were

DEVELOPMENT OF DEMOGRAPHY 103

obtained; such livebirth histories were to be an important feature of the 1946 Family Census. In 1939 M J Elsas and P Moshinsky were taken on by the PIC to work on demographic aspects of housing; a grant was obtained from the Carnegie United Kingdom Trust in support of this work; Elsas later published *Housing Before the War and After* (1st edition 1942, 2nd 1945) and *Housing and the Family* (1947). Meanwhile Glass was working on a revised and enlarged version of his *The Struggle for Population:* this appeared as *Population Policies and Movements in Europe* in 1940.

In 1939 Sir Walter Layton resigned as treasurer of the PIC and L J Cadbury (like Carr-Saunders, Blacker and Glass, an active Eugenics Society member) took over; he remained in this post until 1976. Over the years Cadbury made many, often substantial, contributions to PIC funds and actually paid the research secretary's salary during the strategically very important period 1944-46.

The outbreak of war in 1939 brought the PIC almost to a standstill. Blacker was called up for active service with the Royal Army Medical Corps and was sent overseas; Glass left for the United States to take up a Rockefeller Foundation fellowship. For some time the affairs of the PIC were dealt with by a War Emergency Committee consisting of A M Carr-Saunders, Lord Horder, Sir Charles Close and L J Cadbury. However, from late 1942 onwards the PIC started to be active again - both Blacker and Glass were back in the country by then - and meetings of the full committee began again in January 1943. In November 1944 Glass resumed his duties as full-time research secretary.

In 1943 it was announced that a Royal Commission on Population would be established; the commission actually began its work in 1944. Six members of the PIC were members of the commission or one of its technical committees: C P Blacker, A M Carr-Saunders, D V Glass, H D Henderson,

E Holland and R R Kuczynski. For a while the existence of the Royal Commission actually made it quite difficult for the PIC to develop its own programme of activities since there was the strong possibility that one or other of the PIC's intended projects might be taken over by the commission: indeed PIC members were actively engaged in trying to bring this about. In the event the Royal Commission carried out two very important enquiries both of which essentially emanated from the PIC: the 1946 Family Census and the survey of fertility and contraceptive practice carried out by E Lewis-Faning in 1946-47. The Family Census was based upon a 10% sample of ever married women in Great Britain; livebirth histories were obtained thus making it possible to examine the process of family building in some detail; David Glass directed the enquiry assisted by E Grebenik; the report on the enquiry appeared as their *The Trend and Pattern of Fertility in Great Britain* in 1954. The Lewis-Faning enquiry involved interviews with a substantial sample of ever married women in Great Britain; the survey was not a strictly representative one but it was very important nonetheless as the first ever national survey of birth control practice in Britain; the survey report appeared as Lewis-Faning's *Report on an Enquiry into Family Limitation and Its Influence on Human Fertility during the Past Fifty Years* in 1949.

During the war years the PIC survived financially mainly through a series of grants from the Eugenics Society and a substantial contribution from its own treasurer; in addition it was still using up research funds obtained before the war (including, according to David Glass, private donations amounting to £1766). However, in 1945 the PIC's basic financial situation was transformed when it was awarded a grant by the Nuffield Foundation of £5000 a year for five years, to run from 1 August of 1945. The basic financial security provided by this grant was undoubtedly crucial in enabling the

DEVELOPMENT OF DEMOGRAPHY

PIC to embark on a rather ambitious programme of activities over the next few years.

The first major use to which the Nuffield Foundation money was put was the carrying out of a survey of social and economic aspects of pregnancy and childbirth in 1946. J W B Douglas was director of the enquiry and Griselda Rowntree his assistant. The survey involved interviews by health visitors with all the women in Great Britain who gave birth in a particular week in March 1946 approximately eight weeks after birth; nearly 14,000 women were interviewed. The report on the enquiry, written by Douglas and Rowntree, was published as *Maternity in Great Britain* in 1948.

The 1946 survey has become, by degrees, the MRC National Survey of Health and Development, an extremely important longitudinal study (the Medical Research Council took over in 1962). A large probability sample of the original respondents was re-interviewed in 1948 - the Nuffield Foundation awarded the PIC a special additional grant to make this possible - and this sample has been followed ever since: initially information was collected intermittently from the mothers; later the children became the respondents themselves; in addition a variety of other information about or relevant to respondents was collected on behalf of the study by health visitors, teachers, school doctors and nurses, etc. A huge amount of social, psychological and medical information has been collected for the 1946 birth cohort and a number of books and innumerable articles based upon it: given its longitudinal nature and the very long time-period over which it has operated this can only be regarded as one of the most important surveys in the world.

In 1932 the Scottish Council for Research in Education (SCRE) had administered an intelligence test to all Scottish schoolchildren born in the year 1921. In 1945 the PIC suggested to the Scottish Council that in view of the claim quite often made that the level of national intelligence must have been declining since the working class had more children than the middle class it would be useful to repeat the 1932 survey. It was agreed that the SCRE would repeat the 1932 survey with the PIC finding the money. In 1947 the same intelligence test as had been used in 1932 was administered to all Scottish schoolchildren born in the year 1936 and some educational and demographic information was also collected for these children. The total cost of the enquiry was £5,000: the Eugenics Society gave the PIC a grant of £2000 towards this; the remainder came from the PIC's Nuffield Foundation grant. The first report on the survey was published by the SCRE as *The Trend of Scottish Intelligence* in 1949 (far from suggesting a decline in intelligence the data seemed to indicate the reverse). This was followed by *Social Implications of the 1947 Scottish Mental Survey* in 1953 and *Educational and Other Aspects of the 1947 Scottish Mental Survey* in 1958, both written by J Maxwell and published by the SCRE. In addition to the 1947 Scottish survey the PIC also funded some analogous work in England: this was reported on by W G Emmett in his 'The trend of intelligence in certain districts of England' in *Population Studies* in 1950.

In 1949 the Social Research Division at the LSE together with the Ministry of Labour carried out a sample survey in which more than 9,000 persons 18+ living in Great Britain were interviewed: from the point of view of the Social Research Division the important thing was that the survey provided information on social mobility. The report on this project appeared as *Social Mobility in Britain* in 1954: David Glass was both the editor of this volume and a major contributor to it. The PIC paid part of the cost of the enquiry on the understanding that questions would be included which would permit an examination of the connection between social mobility and fertility. J Berent was taken on by the PIC in 1950 to work on this topic and duly published 'Fertility and social mobility' in *Population Studies* in 1952; he also contributed a chapter on 'Social mobility and marriage' to *Social Mobility in*

Britain. (Berent also wrote 'Relationship between family sizes of two successive generations' which was based on data from the Lewis-Faning survey: this appeared in the Milbank Memorial Fund Quarterly in 1953.) The PIC was subsequently involved in further work on the connection between social mobility and fertility when it met part of the cost of a survey of schoolteachers carried out at the University of London Institute of Education under the direction of J E Floud: two articles by W Scott based on this material were published in *Population Studies:* 'The fertility of teachers in England and Wales' in 1957 and 'Fertility and social mobility among teachers' in 1958.

These, then, were the main activities of the PIC in its early years. As time went on the PIC was to develop a very broad interest in matters of population. However in thinking about its early work both its original concern with low fertility and the fact that many of its most active members were Eugenics Society members (including of course Carr-Saunders, Blacker and Glass) should very much be kept in mind. When the PIC sponsored work on the problems of housing or the costs of education it was largely because these were potential constraints on fertility. The original motivation behind the survey of the births that took place in a particular week in 1946 in Great Britain - now the MRC National Survey of Health and Development - was to investigate the costs and difficulties of, and facilities provided for, maternity. When the PIC lobbied for better official statistics it most of all had improvements in the analysis of fertility in mind. The Family Census of 1946 and the survey of contraceptive practice carried out by Lewis-Faning in 1946-47 (both formally part of the work of the Royal Commission on Population but in many ways arising essentially from the PIC) also represented attempts to obtain an improved picture of the process of family formation. Interest in social class differentials in fertility at least partly stemmed from the pre-occupation of eugenists with the suggestion that the less fit were outbreeding the more fit; similarly work on the connection between fertility and social mobility arose in part from the suggestion that the upwardly mobile who presumably had desirable characteristics were being rendered relatively infertile by their own success. The survey of intelligence carried out in 1947 among Scottish schoolchildren along the same lines as one carried out in 1932 was a conscious attempt to test the proposition that the level of national intelligence was declining.

Postscript

A number of loose ends remain. It is clear that the Eugenics Society was very much involved in the establishment and development of the IPU and its British component, the British Population Society. Why then was this latter organisation overlooked as the independent research body - or at least a framework for it - that the Eugenics Society wanted to see when it set out instead to establish the PIC? Constitutionally, the BPS was apparently very well suited to this role. It occurs to me that the attraction of the PIC may well have been that it was a more or less private body, unlike the BPS, and established its own membership, thus making it possible easily to exclude those with unwelcome political associations or perhaps too 'negative' in their eugenics. One prominent member of both the BPS and the Eugenics Society, G H L F Pitt-Rivers, certainly complained bitterly about the establishment of the PIC which he saw as a rival organisation to the British Population Society (Pitt-Rivers, 1937). Pitt-Rivers was interned by the British authorities during the second world war as a supposed German sympathiser. Carr-Saunders *et al.* would probably not have found it easy either to cope with the enthusiasms of R Ruggles Gates, also a prominent member of both the BPS and the Eugenics Society, who thought it '-probable that all coloured people who show ability in Western civilization have derived it from their white ancestry' (Gates, 1934, p.29).

DEVELOPMENT OF DEMOGRAPHY

The publication of the IPU journal *Population* was discontinued during the war and not revived afterwards. In a sense its place was taken by two non-IPU journals, *Population* in France and *Population Studies* in Britain (the latter published by the PIC), both of which first appeared soon after the war. At least one person did not especially mourn its passing: in a letter to Mauco, then General Secretary of the IPU, dated 6 May 1947, David Glass wrote 'The quality of the material published - was not of such a standard or of such interest - to suggest any real urgency in reviving publication ...' (letter in PIC files).

During the war, albeit with some difficulty, the PIC continued to function whereas the BPS apparently did not. A memorandum, included in the PIC minute book, dated 24 December 1942, drawn up by the 'War Emergency Committee' of the PIC (the memorandum was signed by Sir Charles Arden-Close, C P Blacker, L J Cadbury, A M Carr-Saunders and Lord Horder) included a section under the heading 'The British Population Society', as follows: 'The outbreak of war caused a necessary curtailment of the functions of this Society, and it was decided by the Committee of the International Union to suspend further publication of "Population" after Volume 3 No. 1, which was produced in 1939'. Quite how these matters were part of the PIC's province was not clear. After the war there were discussions between the PIC and the BPS about a possible merger between the two or the PIC taking over the functions of the BPS. In the event the latter course was decided on: in 1947 the BPS was dissolved and absorbed by the PIC which took over its assets (£95 6s 7d) and its obligations (these fortunately turned out to be zero), though not its membership. The PIC thus effectively became for a time the British National Committee of the IPU and represented Britain in the discussions which led to the reorganisation of the IPU and the emergence of the IUSSP.

Note on sources

The sources of information on which this account is based are as follows:

i. The annual reports, minute books and administrative records of the PIC.

ii. The papers of the Eugenics Society (kept in the Contemporary Medical Archives Centre, Wellcome Institute for the History of Medicine, 183 Euston Road, London).

iii. Diaries kept by Sir Charles Close during his attendance at the IPU conferences in Berlin in 1935 and Paris in 1937 (kindly shown to me by his son, Colonel R F Arden-Close).

iv. The published and unpublished materials I have referred to in the account itself.

v. My own earlier work, published in 1988 by the PIC, *The Population Investigation Committee: a Concise History to Mark Its Fiftieth Anniversary*.

References:

Beveridge, W. (1943) Eugenic aspects of children's allowances. *The Eugenics Review* 34, 117.

Blacker, C.P. (1952) *Eugenics*. Gerald Duckworth and Co., London.

Federici, N. (1984) *Italy's Participation in IUSSP during the Early Period of Its Foundation (1928-31); the Italian National Committee and the Congress of Rome in 1931*. Unpublished note prepared for the Working Group on the History of the IUSSP.

Gates, R.R. (1934) Racial and social problems in the light of heredity. *Population* 1,25.

Gini, C. (1933-1935) *Proceedings of the International Congress for Studies on Population, Rome, 1931*. Ten volumes and an index volume. Comitato Italiano per lo Studio dei Problemi della Popolazione, Rome.

Glass, D.V. (1935) The Berlin Population Congress and recent population movements in Germany. *The Eugenics Review* 27,207.

Harmsen, H. and Lohse, F. (1936) *Bevolkerungsfragen*. J.F. Lehmanns Verlag, Munich.

DEVELOPMENT OF DEMOGRAPHY

I.P.U. (1932) *The International Union for the Scientific Investigation of Population Problems (The International Population Union): Its Foundation, Work, Statutes and Regulations.* The Royal Geographical Society, London.

I.P.U. (1938) *Congrès International de la Population, Paris 1937.* Eight volumes. Hermann et Cie., Paris.

I.U.S.S.P. (1985) *The IUSSP in History.* IUSSP, Liège.

Pitt-Rivers, G.H.L.F. (1932) *Problems of Population.* George Allen and Unwin, London.

Pitt-Rivers, G.H.L.F. (1937) *Reports on 1. Progress 2. Science of Population - Methodology and Classification.* Printed booklet prepared for the IPU meeting in Paris in 1937. (Contemporary Medical Archives Centre at the Wellcome Institute for the History of Medicine, London, ref. SA/EUG/D110).

Pitt-Rivers, G.H.L.F. (1938) *The Czech Conspiracy.* Boswell, London.

Sanger, M. (1927) *Proceedings of the World Population Conference.* Edward Arnold and Co., London.

Soloway, R.A. (1990) *Demography and Degeneration.* The University of North Carolina Press, Chapel Hill.

Szreter, S. (1996) *Fertility, Class and Gender in Britain, 1860-1940.* Cambridge University Press, Cambridge.

Human Genetics

John Timson

Human genetics and eugenics have at times been regarded as almost synonymous, particularly in the early decades of this century. The founders of the Eugenics Education Society saw the application of human genetics as a way of solving some of the social problems of their time. They were, perhaps, rather over-optimistic since 90 years ago the science of genetics was in its infancy. However, today we may be in sight of having enough knowledge about our own heredity to realise at least some of the early eugenicists' aims ... if we choose to do so. To appreciate just how far our understanding of the genetics of our species has advanced in the last 90 years it is necessary, briefly, to go back to the beginning.

Around 15,000 to 12,000 years ago the carbon dioxide content of the Earth's atmosphere rose by about 35 per cent[1]. As a result the productivity of plants rose by up to 50 per cent making agriculture economically feasible for the first time. Our ancestors were then hunter-gatherers and many remained so, although no doubt finding life easier. Here and there, people began to actively cultivate plants and domesticate animals. Farming had started and from this change in the human way of life arose settlements, towns, cities, and what we are pleased to call civilisation.

However, while the change in the atmosphere was necessary for this to happen it was not sufficient. The other necessary factor was the development of plant and animal breeding. The early breeders must have soon discovered that like begets like... if you breed rabbits you get a lot of rabbits, nothing else. They also discovered that while this is true there was also a significant amount of variation among the animals and plants they were breeding and so, to get the results they wanted, whether it was fatter cattle, faster horses, or a better harvest, they had to breed from the best they had and reject the less useful strains. A good example of what can happen when humans take control of animal breeding is the modern dog. From lapdog to rottweiler our dogs today remain one species, they are interfertile, and are all descended from the wolf.

Once they had some idea of breeding animals and plants to suit their needs it cannot have been long before our ancestors turned their attention to their own breeding. This would be of particular interest to the ruling groups who had made their way to the top and would want to keep their family there. Hence the arranged marriages within these groups, the harems of many eastern rulers, and at the extreme the sibling matings like those practised by the Pharaohs of Ancient Egypt. Later on such selective mating would be widely practised by various economic and social groups from the castes of India to the guilds of merchants and craftsmen in Europe.

All of this breeding, of animals, plants, and humans, was entirely empirical. It worked enough of the time but our ancestors had no real idea of how the desired traits were inherited although it was often believed to be something in the blood. The start of genetics as a science only really began with the work of two men, Francis Galton and Gregor Mendel, who happened to be born in the same year, 1822. Although they came from very different backgrounds and led very different lives they had one important thing in common — they analysed their results in a quantitative manner. Galton often said, "When you can, count" and Mendel counted his peas.

Mendel's seminal work with peas was published in 1866[2] but was hardly noticed until it was rediscovered and confirmed in 1900. Why did the scientific world ignore Mendel's paper? It was, of course, published in a rather local journal but it was quite widely distributed and its priority was at once recognised when his work was repeated. The real reason, it appears, is that in 1866 the controversy started by Darwin and Wallace's theory of evolution was occupying the attention of most biologists. Today, when this theory is generally accepted by all except the few creationists left, it is not easy to imagine the passions aroused by it at the time. Then, however, evolution by natural selection was the subject of a long-running debate involving people from a wide range of disciplines, rather like the current controversy about the inheritance of intelligence. So, in the second half of the 19th century it must have seemed to those who did read Mendel's paper that it contributed nothing to what they saw as the big issue of the day.

Galton had not heard of Mendel's work when he published his *Hereditary Genius* in 1869[3], his *Inquiries into Human Faculty* in 1883[4], and his *Natural Inheritance* in 1889[5]. If he had known of Mendel's results the history of human genetics might well have been quite different. However, Galton did establish that human traits are inherited, a considerable achievement at the time given the inherent difficulties in studying human genetics. The problem is that humans are perhaps the most unsatisfactory of all organisms for genetic study. The time between successive generations is long, individual families are almost always too small to establish ratios within them, and the test matings necessary for straightforward genetic investigation are not possible. Clearly no geneticist would waste his or her time on such a difficult species were it not for the importance we humans attach to almost any information about ourselves. All the human geneticist then could do was to gather the data available from unplanned experiments in breeding and hope to find a pattern in the data. It was Galton's genius that he was able to do this.

So, if Mendel is to be credited with laying the foundations of modern genetics Galton can be seen as the founder of human genetics. It is remarkable how many of the methods he pioneered are still in use today, for example twin studies, pedigree collection and analysis, and the use of statistics. What is perhaps even more remarkable about Galton's work on human inheritance is the sheer volume of data he collected and presented in his books. In this he was like his cousin, Charles Darwin, who backed up his theory of evolution with an immense amount of data.

Galton was one of Darwin's first supporters and his work on human heredity can be seen as complimentary to Darwin's. Many of our Victorian ancestors were unable to accept either Darwin's or Galton's ideas. To a species which had long regarded itself as specially created to rule over the natural world the ideas that (a) it is only an evolved ape, and (b) that it is subject to the same rules of inheritance as its dogs, farm animals, and crop plants, were hard to accept. Only two centuries earlier we had reluctantly accepted that our planet is not the centre of the universe which had seemed obvious for so long. Now, in the late 19th century it seemed that humans were just one more species of animal. As a species we have always found it hard to believe that we are not, in some way at least, unique.

Naturally much attention in human genetics has been focused on those medical conditions which have a genetic basis. Archibold Garrod was one of the pioneers of medical genetics. His study of what he called inborn errors of metabolism, inherited disorders of body biochemistry such as alkaptonuria and phenylketonuria (PKU), which he published in 1909[6] is important since it can be regarded not only as the starting point for medical genetics but also for biochemical genetics.

Garrod's work, like that of Mendel, was largely overlooked and did not enter the mainstream of genetics until the 1940s. This was partly because many at the time believed that the kind of disorders he studied were caused by an infective agent, probably a bacterium. The direct link between the genes and an organism's biochemistry was only firmly established as a result of Beadle and Tatum's work with the fungus *Neurospora* published in 1941[7]. By reversing the then usual procedure of attempting to work out the chemical basis of known genetic characters they were able to show conclusively that genes controlled known biochemical reactions. This, of course, is precisely what Garrod had inferred over 30 years earlier.

Later, in 1931, Garrod wrote about genes predisposing to disease[8]. This is a more complex situation than simple Mendelian inheritance and it is useful to consider the relative contributions the genes and the environment make to a disorder or a behavioural trait as a spectrum. At one end there are those conditions which are effectively 100 per cent genetic such as PKU and Huntingdon's disease. Individuals with these genes will develop the condition whatever their environment. At the other end of the spectrum are the purely environmental conditions in which the genes play no part such as road traffic accidents. In between are the disorders and traits in which the genes and the environmental interact to produce the phenotype. They are usually known as multifactorial conditions.

Among these multifactorial conditions are many common disorders such as spina bifida, at least some cardiovascular conditions, and possibly Alzheimers. Also in this part of the spectrum are most of the traits of interest to behavioural geneticists such as intelligence and memory, activity level, and sociability[9]. To develop one or more of these conditions, good or bad, a person must carry the predisposing gene or genes and also encounter the right or wrong environmental factor or factors. A good example, one Galton studied, is the height of an adult. There are genes which determine the maximum height a person could reach but their actual achieved height may well depend on nutrition in childhood. However, a pygmy will not grow to six feet tall in any environment.

The four decades between 1910 and 1950 were a time of great advances in our understanding of the mechanisms of inheritance. It was then that many of the basic concepts of genetics were established, most of them with organisms which could be bred and studied in the laboratory, in particular the fruit fly, *Drosophila*, and the fungus, *Neurospora*. Other work at that time concentrated on crop plants especially maize. Genes were shown to be present on the chromosomes in a linear manner which enabled linkage maps to be made and the one-to-one relationship between genes and their protein products was established.

While all this was happening pedigree collection and analysis was the main method of studying human heredity[10]. Pedigrees were felt to provide the raw data of human genetics whether the mode of inheritance was simple Mendelian or multifactorial. They demonstrated, at least when carefully collected, that something, good or bad, was being passed down the generations. Indeed for some of those involved at the time this was the most important aspect of pedigrees and the exact mode of inheritance was of lesser interest. For a time there was a quite serious division of opinion among human geneticists, between those who sought to explain all human inheritance in Mendelian terms and those who sought the answers in statistics.

The Eugenics Education Society, in its wisdom, took no position on this being quite prepared to have speakers from both camps in its education programmes. This was fortunate since, as we now know, the dispute arose from a lack of knowledge at the time and is more apparent than real. In the history of human genetics the importance of the Eugenics Education Society is that it was quite clear that the reason for its existence was to study human heredity and to inform the public of its vital role in human affairs. A quote from its policy pamphlet of 1922-23[11] makes this point well:

"Eugenics does not deny the great importance of environment as a factor in progress, but this is the only society in this country which emphasises the importance of heredity."

Today some of the methodology used by the Eugenics Education Society seems outdated and perhaps, in parts, open to question. No doubt some of its public statements would not now be regarded as politically correct. However, whatever one's views on these matters the fact remains that if the Eugenics Education Society and its successor, the Eugenics Society, had not existed, little, if any, work on human heredity would have been done in Britain at that time. Since the example of the Eugenics Education Society led to the formation of similar organisations in other countries it can perhaps take some of the credit for stimulating the study of human genetics in those countries as well.

The Eugenics Education Society was certainly not taking an easy option. The study of human genetics in this period was difficult and apart from pedigree studies and some work on inherited diseases many geneticists then, for good and obvious reasons, preferred to work on plants and fruit flies. In particular human genetics suffered from an unfortunate and misleading mistake. In 1923 Painter published the human chromosome number as $2n = 48$[12] and this was accepted and taught as correct for some 30 years. The mistake happened because it is technically much easier to stain and count plant chromosomes and the giant salivary gland chromosomes of the fruit fly than those of humans. It was not until 1956[13] that new techniques were developed which revealed that the true human chromosome number is $2n = 46$, that is 22 pairs of autosomes and the sex chromosomes, X and Y in the male and 2X in the female

The ability to visualise and accurately count human chromosomes provided a considerable impetus to the study of human genetics. Down's syndrome had not been regarded as a genetic disorder but chromosome analysis showed that patients suffering from it have 47 chromosomes, an additional No.21, one of the smallest. Some kinds of infertility were also found to be chromosome disorders notably Turner's syndrome in which females have only one X chromosome and Klinefelter's syndrome, males with two Xs and a Y. In time this led to the prenatal diagnosis of Down's syndrome and the possibility of selective termination.

Later developments of better staining techniques enabled geneticists to band and map human chromosomes rather like those of the fruit fly many years earlier and linkage maps directly relating to the chromosomes became possible. Advances in biochemistry led to genetic fingerprinting and ultimately to the Human Genome Project. They also led to the development of laboratory tests for specific genetic conditions. For much of its history progress in human genetics depended to a large extent on the extrapolation of mechanisms of inheritance discovered in other organisms to human data. This has changed radically in recent years and now advances in human genetics are often applied to other organisms. Genetic fingerprinting, developed to identify individual humans, has been used, among other applications, to determine paternity in birds.

In his *Eugenics: Galton and After* published in 1952[14] C P Blacker asked "Should we, indeed, postpone taking action till the human chromosomes have been mapped like those of the fruit fly?" He concluded, of course, that there was much that could be done without waiting on this event and, in any case, he doubted that it would necessarily produce much of practical value. When he wrote this human genetics did not have anything like the predictive value it has today. Probably no one at that time could have foreseen the way human genetics would develop as a result of technical advances which enable us to explore the human chromosomes in far greater detail than those of the fruit fly in Blacker's day.

Today we await the full results of the Human Genome Project (HGP)[15] started in the mid 1980s. Although it is estimated that at least $4 billion will be spent on it and it will undoubtedly provide a wealth of new information about our chromosomes I suspect that Blacker's pessimism of 45 years ago may have been well founded. Locating the gene for a particular disorder and determining its chemical structure does not of itself lead to a cure or even an effective treatment. There is a danger that unrealisable hopes will be raised and suggestions that genetic disorders will all be conquered in the early 21st century seem over-optimistic.

However, laboratory tests for some genetic disorders are already available and more will almost certainly be developed in the years ahead quite possibly as a result of the HGP. These tests have a high predictive value about the future health of an individual although they are not 100 per cent accurate. Not surprisingly they have aroused some public concern and comment both in academic journals[16] and the press[17]. If a person in a family known to suffer from an inherited disorder is tested and found to carry the gene or genes which make it likely that they will not live to retirement should this knowledge be allowed to affect their employment prospects and/or their access to life insurance?

At present there are relatively few such tests and only a few people are likely to be affected by them. However, looking into the future one can imagine a whole range of genetic tests such that a person having none of the wrong genes need only insure himself against death by accident. Those who the tests show to be poor insurance risks, however, may find themselves either unable to take out life insurance or be required to pay much higher premiums. What actually happens will depend on three things:

1. Will insurance companies demand that new clients take the tests?
2. How expensive will they be?
3. Will there be legal restraints on the uses of genetic tests?

Insurance companies are clearly interested in genetic tests which might be seen as simply an extension of the information some already collect on their clients' families. The Association of British Insurers (ABI) issued a policy statement in February 1997[18]. At present they will not ask people to take tests when applying for life insurance. Until March 1999 they will not take into account the results of any genetic tests for applicants for life insurance up to £100,000 linked to a mortgage. For other new applicants individual insurance companies will decide whether to use genetic data or not. Some have already said they will not do so. Meanwhile the ABI has set up a Genetics Committee which will keep the situation under review and draw up a Code of Practice.

Genetic testing is not cheap and for a time at least the expense may limit its use for insurance purposes. Whether or not mass population screening for genetic defects happens in the future must depend on the technology being developed and on the level of demand for such tests. The two are probably linked. A high demand, which might one day be insurance-related, will almost certainly lead to the necessary technology being developed. Otherwise probably only those with a known, serious genetic disorder in their family may be required to take the appropriate test.

Can legislation prevent insurance companies (or employers) using genetic test data? This is already banned in Belgium, Norway, Austria, and at least 13 American states. France has imposed a five-year moratorium on their use while the Dutch allow their use only in regard to policies worth over £70,000[19]. In Britain the Human Genetics Advisory Committee is currently studying the situation[20]. Banning the use of genetic test data may seem a fair way of spreading the risk over all those with life insurance. However, since non-smokers can already insure themselves for less than smokers why should the genetically fit not pay less than those who are less fit? There could well be consumer demand for this. As the ABI statement says much will depend on the way in which genetic testing develops and on public attitudes towards it. Over-regulation by governments could lead to insurance companies operating from offshore islands and by so doing defeating the intention of any legislation.

While the uses of genetic tests are still being debated in the Western countries the People's Republic of China has plans to use them on a large scale and has appointed a French company, Genset, to screen China's 1.2 billion people for serious genetic disorders[21]. This action has outraged those who mistrust the Chinese Government's intentions with regard to the use of such data because of the Law on Maternal and Infant Health Care enacted by that government on 1 June 1995.

Article 1 of this law states "This law is formulated in accordance with the Constitution with a view to ensuring the health of mothers and infants and to improving the quality of the newborn population"[22]. Article 10 states that when a prospective parent is diagnosed as having a serious hereditary disease "inappropriate for child-bearing" the couple can only marry if they undertake long-lasting contraception or are sterilised. Article 18 is concerned with the termination of pregnancies if there is a prenatal diagnosis of "genetic disease of a serious nature."

Although Article 19 states that sterilisation or abortion can only be performed with the signed consent of the prospective parents or their legal guardians it is hardly surprising, given the present Chinese Government's record on human rights, that concern has been expressed about the actual application of this law. However, it is unfortunate that much of this criticism has taken the form of labelling the law 'eugenic' and, therefore, and without further thought, wrong. Much will depend on whether or not the right of an individual to refuse sterilisation or abortion, which Article 19 implies, happens in practice. Time alone will tell if this law is used to improve the quality of the newborn population in a humane manner or for more dubious political purposes.

As a result of the controversy over the Chinese law the Genetical Society following a ballot of its members, in which seven per cent voted, has suspended its membership of the International Genetics Federation because the IGF is organising the 18th International Congress of Genetics in Beijing in 1998[23]. This law has also, in part, led the American Society of Human Genetics to set up a Eugenics Subcommittee of its Social Issues Committee[24]. This will consider the history of eugenics, whether current practices in human genetics are eugenic in nature, and future developments in human genetics which could be used for eugenic purposes.

Clearly there has been a re-awakening of interest in eugenics which is likely to increase as we learn more about our own heredity. If the issues are discussed in a rational manner this would be helpful in leading to a well-informed public debate on the uses of human genetic information. Unfortunately it is only too possible that many will approach the subject with preconceived, anti-eugenic views. Partly this is the result of the distorted misuse of human genetics in the 1930s especially in Nazi Germany. Partly, however, it also stems from the idea that humans should be created equal, even though we quite clearly show the kind of variation found in other organisms. There are still some people who, like our Victorian ancestors, prefer to believe that in some way we are unique and even that to intervene in our own heredity is to interfere with God's plan[25]. For this and other, often personal, reasons some people will, given the choice, decline to take part in any kind of genetic testing.

On the other hand there will be those who will wish to make use of the advances in human genetics when planning their family. Some parents have already shown an interest in choosing the sex of their child and most parents, given the choice, would prefer a healthy baby. At present prenatal diagnosis makes it possible for parents to choose not to have a child with certain disorders such as Down's syndrome. In the future it is at least possible that they will have the opportunity to have what have been called 'designer babies', children likely to be near the parents' ideal child in health, intelligence, athletic ability, and even looks.

This, if it happens, would amount to individual eugenic choice in order to give a child the best possible start in life in a society in which hereditary traits have become important in the social hierarchy. An editorial in *Trends in Biotechnology*[26] in 1989 included the statement "Human improvement is a fact of life, not because of the state eugenics committee, but because of consumer demand." Although there may well be considerable consumer demand there is also powerful opposition to any attempt to improve the human gene pool, even by individuals, from some religious groups. In the long run consumer demand is likely to triumph, at least in liberal democracies, but the controversy over the uses of human genetic knowledge will be one of the most important social issues under discussion in the years to come.

Genetics and eugenics have been part of the human experience for several thousand years even though the words themselves are of quite recent origin. For almost all of that time trial-and-error has been the basis on which breeders worked and it has only been in this century that we have begun to understand the laws of inheritance and their biochemical basis. For much of the early history of modern genetics human genetics has been the Cinderella of the science. During that time few outside the eugenics movement contributed much to the study of human heredity. Today human genetics has a much higher profile and is contributing to genetics as a whole in a way rarely seen in the past.

Our use of genetics until quite recently has largely been to increase both the quality and quantity of our crop plants and domesticated animals. This has allowed us to increase the quantity but rarely the quality of our own species and is a major cause of the population explosion. In recent years genetic counselling and prenatal diagnosis of inherited disorders have in a small way improved the quality of human life for a few people. However, much of the world's population continues to breed too rapidly in happy ignorance of genetics and the problems they are bequeathing to their descendants.

The HGP could lead in time to effective treatments for some genetic disorders. This possibility brings with it the danger that the genes responsible are then more likely to be passed on. Clearly it would be wrong to deny patients treatment but it would be equally wrong to encourage these patients to reproduce and they should be persuaded whenever possible not to do so. Genetic disorders can never be totally eliminated since new mutations arise in every generation. However, the knowledge we have now and are likely to have in the fairly near future should enable us to reduce considerably the number of people suffering from genetic disorders. How we use this knowledge will, I believe, have a profound effect on the future of the human species. We would do well, I suggest, to remember the old adage that prevention is better than cure.

Acknowledgements: I would like to thank Dr John Peel for the loan of books which have made writing this much easier and the Association of British Insurers for information about genetics and insurance.

References:

[1] Rowan F Sage, "Was Low Atmospheric CO_2 during the Pleistocene a Limiting Factor for the Origin of Agriculture?" *Global Change Biology*, (1995), Vol 1, 93-106.

[2] Gregor Mendel, "Versuche über Pflanzenhybriden", *Verh. naturf. Ver Brünn*, (1965), Vol 10, 1.

[3] Francis Galton, *Hereditary Genius*, Macmillan, London, 1869.

[4] Francis Galton, *Inquiries into Human Faculty*, Macmillan, London, 1883.

[5] Francis Galton, *Natural Inheritance*, Macmillan, London, 1889.

[6] Archibold E Garrod, *Inborn Errors of Metabolism*, Frowde, Hodder, and Stoughton, London, 1909.

[7] G W Beadle and E L Tatum, "Genetic Control of Biochemical Reactions in *Neurospora*", *Proc. Nat. Acad. Sci*, (1941) Vol 27, 499-506.

[8] Archibold E Garrod, *The Inborn Factors in Disease*, Clarendon Press, Oxford, 1931.

[9] Stephanie L Sherman et al., "Recent Developments in Human Behavioral Genetics: Past Accomplishments and Future Directions", *Am. J. Hum. Genet.*, (1977), Vol 60, 1265-1275.

[10] Pauline M H Mazumdar, *Eugenics, Human Genetics and Human Failings*, Routledge, London and New York, 1992.

[11] Eugenics Education Society, Policy Pamphlet, 1922-23, Eugenics Society Papers, C108.

[12] T S Painter, "Studies in Mammalian Spermatogenesis II. The Spermatogenesis in Man", *J. Expt. Zool.*, (1923), Vol 37, 291.

[13] J H Tjio and A Levan, "The Chromosome Number of Man", *Hereditas*, (1956), Vol 42, 1, and C E Ford and J L Hamerton, "The Chromosomes of Man", *Nature*, (1956), Vol 175, 1020.

[14] Charles P Blacker, *Eugenics: Galton and After*, Duckworth, London 1952.

[15] Daniel J. Kevles and Leroy Hood (Eds), *The Code of Codes*, Harvard University Press, Cambridge, Mass., and London, 1992, and John Timson, "The Human Genome Project ... for Better or Worse?" *Galton Institute Newsletter*, No. 14, September 1994.

[16] H Ostrer et al., "Insurance and Genetic Testing: Where are we now?", *Am. J. Hum. Genet.*, (1993), Vol 52, 565; John Timson, "Goodbye to Life Insurance?", *New Scientist*, 14 August, 1993; John Timson, "Genetic Testing

and Insurance", *Galton Institute Newsletter*, No. 10, 1993; Alison Motluk, "Your Money or Your Life", *New Scientist*, 18 January 1997; Sandy Raeburn, "The Big, the Small, and the Average", *New Scientist*, 3 May 1997; Helen Middleton-Price, "Behind the Screening", *Science and Public Affairs*, Spring 1997; Robert J Pokorski, "Insurance Underwriting in the Genetic Era", *Am. J. Hum. Genet.* (1997) Vol 60, 205.

[17] *Sunday Times*, 13 October 1996; *The Times*, 3 February 1997, 19 February 1997, 22 February 1997, 24 February 1997.

[18] Association of British Insurers, *Life Insurance and Genetics, A Policy Statement*, 1997.

[19] *The Times*, 17 June 1997.

[20] *The Times*, 10 July 1997.

[21] Andy Coghlan, "Chinese Deal Sparks Eugenics protests", *New Scientist*, 16 November 1996.

[22] *Genetical Society Newsletter*, No. 30, July 1996.

[23] *Genetical Society Newsletter*, No. 32, December 1996.

[24] American Society of Human genetics, *News*, March 1997.

[25] Ellen W Clayton et al., "Lack of Interest by Nonpregnant Couples in Population-Based Cystic Fibrosis Carrier Screening", *Am. J. Hum. Genet.* (1996), Vol 58, 617, and John Timson, "Genetic Testing and God's Plan", *Galton Institute Newsletter*, No 22, September 1996.

[26] Editorial, "Geneticism and Freedom of Choice", *Trends in Biotechnology*, September 1989, 221.

Ninety Years of Psychometrics

Paul Kline

In this paper I shall not attempt to provide a detailed history of psychometrics which would, in my view be inordinately long and tedious. Rather I shall highlight key points in its development.

Galton, of course, in Victorian England, was one of the first scientists systematically to investigate individual differences. Psychometrics, which is usually defined as the quantitative study of such differences by means of tests, was essentially his invention although Galton's interests ranged far more widely than this. He was a scientist rather than a psychologist or psychometrist.

Binet is usually regarded as the *fons et origo* of intelligence testing in its modern form. With Simon he devised the first standardised intelligence test – the Binet Scale – an individual rather than a group test. This Binet scale was intended to discover which children were likely to benefit from education rather than as a measure of the variable intelligence. However in the *Experimental Study of Intelligence* (Binet, 1903) it is clear that Binet rightly considered that his test was measuring intelligence. He selected items for his test if they demonstrated the proper age increments. Thus for Binet a good item was one which was answered correctly by a higher proportion of each older age group until at a certain age all would get it right. This was because on theoretical grounds he believed that human abilities steadily increase with age until about the age of sixteen.

It was this age-based approach to the selection of items that led Binet to develop the notion of the intelligence quotient, the IQ, which he calculated as mental age (measured by the test) divided by chronological age multiplied by 100. Thus a child of 10 who can solve the items designed for children of fourteen years has an IQ of 140. Even today a modified form of this IQ is still in use.

There are three points to be noted about this work of Binet. The first is that the Simon-Binet test still exists today, in modern guise, but still recognisable, as the Stanford-Binet - a revision that was carried out by Terman. This is surely remarkable that a test devised nearly 100 years ago should still be used. It is also depressing that in this century psychometrics has apparently advanced so little.

The second point worthy of note is that despite its immense longevity and despite the fact that Binet is referred to in many textbooks as the father of mental testing, this work has turned out to be a byway, not at all central to the development of psychometrics. However Terman's studies of genius deserve mention. He selected, using the revised test, gifted children aged four and a control group and followed them through their lives. The performance of the group selected for high IQ was vastly superior on almost any variable, likely to be related to intelligence. This was powerful support for the efficacy of the test and the importance of the general reasoning factor (Terman and Oden 1959).

This leads me to the third point. The reason that the work of Binet has been relatively uninfluential in the history of psychometrics can be attributed to the fact that one year later, here in London, at University College, Spearman published what is probably the seminal paper in psychometrics: "General Intelligence: objectively determined and measured". It was seminal for two reasons. In the first place it introduced the use of factor analysis to the study of individual differences. Secondly it also introduced the notion of g, the general intelligence factor. Like the work of Binet, this work of Spearman has lived on. Factor analysis is still the typical method of modern psychometrics even though it has been developed extensively especially since the advent of computing. Furthermore g is still a viable concept. The most recent and authoritative study of ability factors, that of Carroll (1993), indicates that g is a mixture of two factors, fluid and crystallised ability, a combination, it should be added, measured by many modern intelligence tests. It is from this pioneering study that psychometrics has developed. The modern psychometrician factor analyses large batteries of tests and then attempts to identify the resulting factors. This is exactly what Spearman did in all his work. Indeed Spearman founded what has been called the London School which was hugely influential not only in psychometrics but in psychology generally, especially in the applied fields of educational and occupational psychology. Even as I am speaking now in many rooms here in London hapless applicants for jobs will be filling in tests and questionnaires and many of these will be directly traceable to or even developed by members of the London School.

It is a tragedy for psychology that this London School has almost disappeared swept away by the tide of egalitarianism and the rise of innumerate social sciences in our universities. I shall now turn to this London School and try to illustrate the importance of their work to psychometrics.

Spearman spent his career in the study of g, general intelligence. He claimed that performance in any problem solving field was attributable to g plus a further ability specific to that particular field. This aspect of the work has not stood the test of time. As we shall see, rather than many specific factors there are broader and more general factors of ability. However it was not so much the findings as the methods which have made Spearman so powerful an influence on psychometrics. His use of factor analysis opened up huge possibilities for research, ones that were brilliantly realised by the London School and taken up elsewhere.

Before I discuss the work of the London School, I need to say something about factor analysis. Factor analysis is a method of simplifying complex sets of data. An example will clarify the method. One of Spearman's interests was understanding school achievement. Why is it the case that performance in school subjects is positively correlated, that the person good at maths is good at English and so on? To answer this question Spearman applied a form of factor analysis, the method of triads, to the correlations. A factor is simply a summation of variables and factor analysis tries to account for the correlations between variables in terms of a smaller number of factors. Each variable, in this case the different school subjects, correlates with these factors. A factor is identified initially from these correlations with the variables. Spearman found a general factor, that is one which correlated with all the subjects. This was his g, the general reasoning factor of general intelligence. He called it general because all subjects were correlated with it.

G correlates more highly with some subjects than others. In a recent study of intelligence, carried out by one of my students, we found that the g factor loaded highly on science subjects but less so on social sciences and sociology. This should not be a cause of surprise and many years ago Alice Heim studying the same g factor among faculties in Cambridge University showed that the lowest scorers were the medical and education students. Some subjects demand hard problem solving and others are but feats of tedious recall. It really means that some subjects are hard and others easy.

The importance of factor analysis as a statistical technique lies in its ability to pull out from complex data a small number of factors which account for much of the variance. Thus in the study of individual differences in ability factor analysis has identified five factors which account for about 70% of the variance (variations in human abilities). These are fluid and crystallised ability, fluency, cognitive speed and visualisation.

To understand abilities, therefore, it is these factors which should be studied. It should also be pointed out that it has been shown by Cattell (1978) that frequently factors have a causal status. Much of the history of psychometrics has consisted of the application of factor analysis to new areas of psychology and to the refinement of factor analysis as a mathematical technique. This has been necessary because it so happens that there is an infinite number of mathematically equivalent factor analyses. How do you choose? That is the question. This problem alone apart from the mathematical complexity led many psychologists to ignore the method and this is particularly true of the University of Cambridge, under Bartlett, whose influence did much to limit the spread of the London School in British psychology. But I jump ahead.

Such was the beginning of modern psychometrics in London at University College, one of the few British contributions to psychology. The history of psychometrics today represents the developments from that date both in this country and in the United States. In relatively recent times important contributions to psychometrics have come from Europe as psychology diverged from its philosophical bases, always stronger in Europe than in the grimly empiricist Great Britain.

To illuminate the history of the subject I shall concentrate on the issues with which it has been largely concerned and discuss how they have developed into modern psychometrics. First the issues. Perhaps it may come as a surprise to note that these are few in number and, as I have hinted already, that they were conceptualised at the beginning of the subject. I shall now list them

1. **The Structure of Abilities.** What are the main ability factors? How may they be developed and nurtured? What are their neural substrates? This was essentially the concern of Spearman. It continues today with special emphasis on the last question.

2. **The Structure of Personality.** Because the factor analysis of abilities had proven successful identical methods were applied to the study of personality both here and in America. In this field the contributions of the London School are particularly impressive and influential. I refer here to the work of Cattell and Eysenck.

3. **The Genetics of Human Ability and Personality.** This began with studies of twins reared apart and together. More recently complex biometric models have been applied to the scores on personality and ability tests of individuals of differing degrees of relatedness. Highly interesting and contra-intuitive findings have been recorded in this area.

4. **The Prediction of Human Behaviour.** There have been two aims in this endeavour. First there has been the theoretical goal of being able to predict human behaviour - a task which has proved itself, so far, beyond the reach of psychometrics. The second more modest aim has been more successful – predicting performance in jobs and education, the use of psychometrics in the applied field. Psychometrics is used in clinical psychology but I have not time, in this lecture, to discuss that aspect of its application.

5. **Statistical Methods.** Since Spearman first introduced factor analysis there have been formidable developments in the statistical methods used by psychometrists. I shall deal only briefly with these since their mathematical complexity is suitable only for specialised audiences.

To discuss the history of psychometrics I shall deal separately with each of these sections and I shall begin with the first, the structure of abilities.

Spearman's work in London was carried on by the great and now notorious psychologist – Sir Cyril Burt. Since the revelation that he had faked his twin data his reputation has entirely collapsed both among psychologists, who ought to know better, and the general public. I do not want to condone cheating which in the Sciences is particularly pernicious. However this should not detract from the fact that Burt was a superb statistician and educational psychologist. He invented the notion of educational psychology and made pioneering efforts to help backward and delinquent children. Finally, as has already been noted, Burt trained a number of well-known psychometrists of whom the most outstanding were Eysenck and Cattell.

Burt made a considerable contribution to the understanding of the structure of human abilities by his efforts to synthesise two disparate views: the two factor theory of Spearman, general ability plus a large number of specifics and an approach which was developed in the thirties and forties by the great American psychometrician – Thurstone - work which was fully reported in Thurstone (1947). This American view denied the importance of general ability. Instead a number of group factors (9) were proposed. These included, for example, spatial ability, abstract thinking, verbal ability, numerical ability, and deductive reasoning. Opponents of factor analysis seized on these two disparate descriptions of human abilities as evidence that the method was of little value since the two leading practitioners were unable to agree. Burt (1955) however was able to demonstrate that these two views were not as different as first appeared.

First he had already modified the work of Spearman by showing that these specific factors were not wholly independent. Thus, for example, it would be strange if the specific factor for learning German were radically different from that for learning any other modern language. Thus Burt showed that the structure of abilities was better explained by g, general ability and a small number of group factors – verbal and mathematical and spatial. The Thurstone model fitted this hierarchical model of Burt because the nine factors were themselves correlated. If these were factor analysed then the general factor and a few group factors emerged. This hierarchical model of Burt held sway until it was revised and improved by Cattell (1971) and polished by Carroll (1993).

This brings me to Raymond Cattell, undoubtedly one of the pivotal figures in Psychometrics. Cattell is one of the London School. He studied chemistry first, in the University of London and then began to work with Burt. Following the master he first founded the educational psychological service in Leicester, all the while carrying out factor analytic researches into ability and also into personality. This work, beginning in the thirties still continues. Most of it was done in Illinois where Cattell became a Distinguished Research Professor. After his retirement his research still goes on at Hawaii although Cattell is now 92. To attempt to summarise his immense contribution to the subject is hard because he has contributed in a major way to all the categories of research which I described previously. He has published around thirty books and about 500 papers in journals and it can be said that none of these is popular or easy to read. Indeed in this country there are few psychologists who are well acquainted with his work.

In the field of abilities Cattell (1971) definitively demonstrated that g, general intelligence comprised two correlated factors – fluid and crystallised ability. The former has a large genetic component in its variance. It might be regarded best as the basic reasoning capacity of the brain determined by the quality of our neural structures. Crystallised intelligence, on the other hand, results from the investment of fluid ability in the skills which are valued in the culture. Thus in Great Britain and the West crystallised intelligence is reflected in performance at most academic subjects and in professional jobs. In other cultures this may not be the case to the same extent. In the early years there is little difference between these forms of g, but as children develop there is divergence, differing according to the stimulation they receive. As has been mentioned Cattell found three other important factors: visualisation, an ability useful in chess and engineering, for example; retrieval or fluency, an ability which is thought to be important in creative endeavours, and finally cognitive speed. These are the most important broad factors.

Carroll (1993) also deserves mention. For many years he was Professor of Education at Harvard. However he went to Chapel Hill, North Carolina in the Laboratory where Thurstone worked and which bears his name. His research on the structure of abilities is remarkable. He has reworked most of the important studies of abilities from their original data. This he did because until recently there has been little agreement concerning the best methods of factor analysis. Thus to all these researches he applied the best methods and his findings were essentially those of Cattell. I say essentially because there can be no definitive list of human abilities, they are simply too many and disparate. For example there is a well-known skill of cheese testing for ripeness, a procedure now replaced by computer controlled machines in modern dairies. However it is a human skill but, as far as I know, there are no tests for it and it appears in no list of factors. To sum up, in the field of the structure of abilities, we have not journeyed far from Spearman and it is fitting to note that a major contribution to our knowledge came from one of the London School.

I shall now examine the development of the study of the structure of personality, a field which began in the late thirties, with the work of Guilford in America and Cattell in the UK, and which actually continues to the present day. Accounts of this early work can be found in Guilford (1959) and Cattell (1957). Until recently, unlike the field of abilities, there was little agreement as to the structure of personality. Indeed the present speaker argued that were as many factor analytic descriptions of personality as there were factor analysts (Kline, 1993). However there is now consensus, although in my view this is misguided. A major cause of all this confusion lies in the fact that, as I have mentioned, it is a difficult decision to choose which of the infinite number of factor analytic solutions is the best. Other reasons are more diffuse but reside in the unfortunate split, in many universities, between psychometrics and psychology.

To understand the history of these studies it is necessary briefly to describe their rationale. Because factor analysis had yielded a successful account of the nature of abilities it was decided by the pioneers in this field to apply the method to personality. However there are obvious problems here since it by no means clear what measures or tests would have to appear in such a study. This is because there are no clear theories of personality. Indeed the same objection, which I raised to the factor analyses of personality, can be raised against theories of personality. Cattell solved this problem by arguing that personality traits should be investigated. If a trait exists there must be a name for it and thus the basis of his work was the factor analysis of traits derived from a dictionary search of the English language. First subjects were rated for all these traits and after factor analysis personality tests to measure the underlying factors were developed. These are the famous Cattell Personality Tests with versions for subjects aged four years and upwards (Cattell and Johnson, 1986). Cattell still argues that his 16 factors constitute the best description of personality, both theoretically and for applied psychology and it has to be said that his personality tests are found useful, even today, in occupational selection. In his many publications Cattell has developed a genuine psychometric psychology based upon his factor analytic variables, both in personality and ability (Cattell, 1981).

Guilford (1897–1987) should be briefly mentioned at this point. He was a most distinguished psychometrist and statistician, working in the University of California for more than fifty years. Although, like Cattell, Guilford produced a series of personality tests these never made the same impact on psychology. This is because Guilford did not attempt to construct a theory of personality although he made a remarkable contribution to the study of human abilities which does deserve mention (Guilford, 1969). Guilford had been shocked by the launch of the Russian sputnik, the first man-made satellite. He argued that Russia had leaped ahead of the USA because the American education system had concentrated on what he called convergent thinking. This is the ability measured by conventional intelligence tests which contain items which have one correct solution. He believed that creativity depended upon divergent thinking, the kind of thinking associated with creativity where the problem to be solved has several solutions and the best has to be selected.

Much of Guilford's work was devoted to the study and assessment of creativity and divergent thinking. He built up a unique structure of abilities which contained 120 abilities and which contrasted convergent and divergent thinking. For many years this was considered to be a brilliant account of the subject and Liam Hudson (1966) made much of it in this country. Unfortunately it has been shown that the factor analytic methods he used were seriously flawed (Horn and Knapp, 1973) which is why I have discussed it here rather than in the previous section.

This brings me to the work of Hans Eysenck, who is the most cited of British psychologists and who has more than 1000 publications to his name. Since the forties he has worked at the Maudsley Hospital here in London and his contribution to psychology in general and to the understanding of the structure of personality and its underlying basis cannot be over-estimated. Eysenck was a refugee from Germany and actually improved his English at my university, the University of Exeter, before proceeding to read psychology at London, completing his PhD with Burt. It was here that he learned of the power of factor analysis but he also realised, as did Cattell, the importance of combining the findings with experimental work.

Whereas Cattell attempted to map the whole of personality, Eysenck concentrated on extraversion, which had been isolated by Guilford, and on neuroticism, a factor which he first observed and measured in the psychiatric patients where he worked. Since that time Eysenck has relentlessly explored the psychological and physiological nature of these two factors, using experimental psychology, physiology, genetics and applied psychology in his work. Much of this was summarised by Eysenck (1967) and later developments can be found described in my brief account (Kline, 1993). In so doing he has isolated a third factor, psychoticism, the final factor in his three-factor model of personality. What are these factors?

Extraversion is a bipolar trait, the low pole being introversion, the high extraversion. It is normally distributed, as is intelligence, meaning that most score around the mean level with only a few extremes. All Eysenck's variables are of this kind. The extravert is noisy, sociable, pushy, interested in the outer world rather than her inner feelings. In the male form they are common in bars and around Rugby grounds. The introvert is the opposite of this, sensitive, concerned with the inner world, quiet, bookish and generally withdrawn. These are simply descriptive terms. The work of Eysenck and his colleagues has gone far beyond this into the psychological and neurological bases of the trait. Extraversion is related to the arousability of the central nervous system. The extravert is lowly aroused – hence her liking for noisy parties. She is stimulus hungry. The introvert, on the other hand, is high on arousal. That is why she longs for solitude, and piece and quiet. She needs little stimulation. Work in applied psychology supports this notion. On dull repetitive tasks the extravert starts well but gets bored and makes errors. The introvert, on the contrary, is able to plough on. Biometric genetic studies have shown that there is a considerable genetic determination in the population variance and that the environmental determinants are unshared, not shared. This means that it is the stimulation, unique to each person rather than what is common to the members of a family which is significant in personality development. This finding which is common to neuroticism and psychoticism applies also to intelligence. Eysenck has also demonstrated that extraverts are less easy to condition than introverts and since, further, conditioning is highly relevant to much social learning, it can be seen that there is considerable psychological significance to this variable.

Neuroticism is essentially trait anxiety. Almost everybody experiences anxiety from time to time so the variable is easy to describe: the highly anxious person worries about everything, has rather rapid mood swings, feels sick or gets headaches when stressed and sweats easily. Highly anxious people are not suited to stressful occupations where there may be threats to life and limb. I must point out that this trait anxiety should be distinguished from state anxiety. This latter is the anxiety felt at any threatening situation. We all feel anxious before exams or doctors' examinations. This is quite normal. Trait anxiety refers to our general resting level of anxiety in our lives. As with extraversion there is a considerable genetic component in the population variance and this is not surprising since it is argued that this trait is related to the lability of the autonomic nervous system. This accounts for the physiological concomitants of anxiety – sweating, rapid heartbeat, horrible feelings in the stomach and, in extremis, fainting. Needless to say neuroticism is implicated in the development of psychiatric disorders, as is extraversion. The hysteric is high on extraversion and neuroticism, the obsessional is high on neuroticism but low on extraversion.

Eysenck's third factor, psychoticism, was fully introduced into his system in 1975 with the development of a new test, the EPQ (Eysenck and Eysenck, 1975). Psychoticism, descriptively, consists of tough-mindedness, lack of empathy, cruelty and liking strange or violent sensations. As might be expected it is higher in males than females and is often raised in criminals. Whereas those high on N, under stress, may develop neurotic symptoms those high on P tend to develop psychoses, mental disorders characterised by lack of contact with reality. Its physiological basis seems to reside in levels of male hormone, hence its elevation in violent criminals and the tendency of women to score lower on the scale. In fact it makes good evolutionary sense from the viewpoint of child rearing if females are more empathic than males, although to say so offends political rectitude. As with the other two factors psychoticism appears to have considerable genetic determination.

Eysenck (1967, 1981) has constructed a powerful theory of personality using these three factors, one which extends into the prediction of conditionability, the susceptibility to mental disorders and psychosomatic illness, including even cancer and heart disease, sexual and marital behaviour, criminality and smoking, to name but a few.

Such brief summaries can hardly do full justice to these two great theorists and researchers from the London School. How are they to be reconciled, three factors or sixteen? In fact reconciliation is not so difficult as it might be thought. The Cattell factors are correlated. If these correlations are themselves subjected to factor analysis these three Eysenck factors emerge, although Cattell refers to neuroticism as anxiety and the psychoticism factor is tough-mindedness. It would appear, therefore, that there is a reasonable consensus.

In psychology consensus is a dirty word, and there is none. One obvious objection to Eysenck's work lies in the reasonable claim that surely the richness of human personality cannot sensibly be resolved into three factors. This intuitive position is supported by more technical research literature since, just for example, the authoritarian personality has been described (Adorno et al, 1951) and achievement motivation is another well-known variable (McClelland, 1961). In fact a new model has become popular in the last ten years – the five-factor model of McCrae and Costa (1990), who are based in America. This model, as the name suggests, proposes five factors: three are essentially the same as those of Eysenck, extraversion, anxiety and agreeableness, which is the low pole of psychoticism. In addition they posit conscientiousness and openness. However although they have demonstrated that these five factors can be obtained by factoring most other questionnaires, the picture is not as clear as is often maintained. For example, in recent studies of their model, using their test, the NEO, conscientiousness and agreeableness were not separate either from each other or from anxiety. There must be doubt whether these five factors are a good account of the factor structure (Kline, 1997).

As was the case with the structure of abilities, work in the field of the structure of personality has made surprisingly little progress in the last fifty years. Actually it can be argued that it will turn out to have been something of a dead end. This is because recent work on the human genome and on brain physiology using Pet scans and other analogous techniques is likely to render questionnaires redundant as serious scientific measures although they may remain convenient for occupational psychology.

I shall now examine the third category of work with which psychometrics has been associated. This is the genetics of ability and personality. This is the field that is closest to the main interests of the Galton Institute. This is also the field which, since the last world war, has attracted to itself the opprobrium of the liberal mind. Over the whole endeavour hangs the ghost of Mengele and it is difficult to conceive of research more horrible than his. More recently, as I have mentioned, the unmasking of Burt has added to the problems. However, in principle, this is a perfectly respectable field of enquiry and I shall summarise its history admitting that it is possible to misuse the findings, as it is in many areas of science.

The first point which I wish to emphasise, because it is frequently forgotten in the heat of debate, is the nature of the question which psychometrics is attempting to answer. It is this: to what extent is the population variance or variability determined by genetic and environmental factors? Note the "and" in genetic and environmental. For personality and ability traits it is inconceivable that there could be an either or in this case. Note again the variance in the population. In a different population the results could well be different. There is a second point. Popular science writers, who are not psychologists, such as Gould (1981), have argued that it is impossible that intelligence could be treated as a variable to be inherited or not on account of its complexity. Both assume that attempts to answer this question are absurd and doomed to failure. However, this is incorrect, the voice of ignorance and prejudice. The variable is a test score on an intelligence test or a personality questionnaire. In principle this is no different from the size of wheat or the length of an oak leaf, variables which are certainly open to the study of their determining factors.

The first approach to this question was to investigate the correlation between pairs of twins. These are ideal subjects since monozygotic twins are genetically identical while dizygotic twins share only half their genes and are genetically no more alike than siblings. From this it clear that any differences between MZ twins must be environmentally determined and that the extent by which MZ twins are more alike than DZ twins of the same sex must reflect their genetic similarity. It is also obvious that a good natural experiment is provided where identical twins are reared apart and where adoption takes place. Here any similarity between child and adoptive mother must be environmentally determined and similarly the correlation between an adopted child and its biological mother must be entirely genetically determined, if the separation was early in life. This was the methodology of the early pioneers of the psychometric study of the genetic basis of intelligence. MZ and DZ twins reared apart and together, and adopted children and their natural and adoptive parents were tested for intelligence, both here and in America.

I shall not describe in any detail the results of this early work for a number of reasons. In the first place the numbers of twins in the studies were small. Secondly it is obvious that it a sample of twins is not representative of singletons. These early findings have been well summarised in Vernon (1979), a notable British psychometrist, and one of the few who were not nurtured up by Burt, although they worked together on a series of educational tests. He showed that there was remarkable agreement and that these kinship correlations, as they are called, for intelligence, are highly similar in pattern to for height and weight. There can be little doubt that it would have to be argued from these data that variation in intelligence had a considerable genetic determination.

Burt (1966) published in the British journal of Psychology what was then the largest study of twins reared apart and together. This appeared to demonstrate that intelligence, as measured by the WISC, had a strong genetic component in its variation in Great Britain. He attributed about 70% of the variation to genetic factors. The sample in this study had been built up, it was claimed, over a number of years and some of these data had been published previously. Yet despite these increases in sample size the correlations quoted remained identical. It was on this basis that it was decided after careful scrutiny that Burt must have fabricated these data, a claim that has not been rebutted. (Hearnshaw, 1979). Yet there are some strange features in this case. First it is unclear why Burt, if he was cheating, quoted the identical correlations. Such identity is impossible, as Burt well knew, far better than his critics. Had he changed them at, say, the second decimal place detection would have been unlikely in the extreme. The fact that these results are suspect has led many non-specialists, quite without logic, to assume that the findings were wrong and that the determination of intelligence is social. This affair has done the cause of psychometrics no good. It has led opponents, and there are many on political grounds, to argue that the area is fraudulent and that, despite this, there are no genetic factors determining human ability.

This twin study method, on its own, is now outmoded. Biometrics, a statistical method developed by Fisher originally for agricultural experiments utilises complex models of heritability, taking into account, dominance and assortative mating, just for example. The data on which these biometric models are based includes the correlations between individuals of all degrees of relatedness. Cattell (1982) also introduced a similar approach, the MAVA method, and these have yielded remarkable findings which ought to have transformed the study of personality and ability. Sadly these are not well known beyond psychometrics. .

The first point is that the weighted average of the correlations between pairs of identical twins reared together and apart, is precisely that invented by Burt. Perhaps of more significance for psychology is that the main ability and personality factors have a considerable genetic component in their variation, ranging from about 70% for intelligence to 50% for some of the personality factors and that the important environmental factors are non-shared. Thus family differences, such as money, education and social class, the variables beloved of sociologists, appear to be uninfluential in the determination of personality and ability.

This modern work in psychometrics is fascinating in its own right and it has forced psychologists to rethink the nature of the environment. It is highly interesting that although there is an almost universal assumption that environmental factors are of great developmental importance just what these factors are and how they operate has never been specified. This biometric work suggests that these influences are subtle in the extreme: differences in the way a mother interacts with different children would be an obvious example. There is a further point of interest. With the gradual unfolding of the genome new genetic data are being revealed. The combination of biometric research with this more traditional genetic approach is certain to prove fruitful and results are now appearing from the Institute of Psychiatry, under the direction of Plomin.

I shall now examine the history and development of psychometrics in applied psychology. It can be said at the outset that without doubt this is the most controversial area. Few people like to have their intelligence put to the test and quite wrongly psychometrics is popularly seen as intelligence testing. In fact, psychometrics has been applied in four areas - educational, occupational and clinical psychology. These constitute the more respectable application. The fourth area, which I shall discuss first and briefly, is viewed by many as quite appalling, a view with which I am in full agreement.

This might be called the social engineering application. It was used in the 1920's and thirties, as has been described by Kamin (1974), for screening immigrants to America. The authorities wished to exclude those of low intelligence whom they felt likely to be a burden on the state. Intelligence tests were administered to these subjects even when their English was poor or non-existent. This was, of course, a profound injustice. It is simply a gross misuse of psychometric testing. Recently it has emerged that in Canada a policy of sterilising orphan girls if they were below a given level of intelligence was in operation from the mid thirties until the early seventies. It is attempts of this kind to control populations based upon intelligence test scores which have brought psychometrics, especially among liberal minded people into considerable disrepute. In opposition to this kind of policy which was pursued also by the Nazis in pre-war Germany, Stalin, in the USSR banned the use of intelligence tests and any other kind of psychometric instrument. I shall say no more about these abuses of psychometrics, other than to note that intelligence tests were not designed for these purposes and are worthless used thus. The phenomenon of regression to the mean and the meiotic reassembly of genes are sufficient to guarantee that such selection would be futile even if it were desirable.

I shall now turn to the history of psychometrics in educational psychology. In Great Britain, at least, it is distinguished history although in the last twenty years or so the involvement in psychometric testing has considerable diminished. In the last few years, testing is back with a bang in schools but it is not, unfortunately, high quality psychometrics. Far too may of the questions, for example, have to be subjectively scored.

Burt was the inventor, single-handed, of educational psychology. Educational psychological services which today can be found in every local educational authority have been built essentially to his pattern. Educational psychologists are called in to help with the education of children with educational problems. It is only the means by which they do this that have changed over the years. Burt used intelligence tests whenever a problem child was identified. His argument was that it was impossible to estimate how well a child ought to progress at school unless one knew her intelligence test score. This makes such good sense that it is a source of amazement to me that intelligence testing is no longer a regular feature of educational psychology. Certainly when I trained as an educational psychologist in Aberdeen in the early 1960's we had to give a large number of supervised intelligence tests and other psychometric instruments. It was work of this kind which led Burt to his excellent publications on backward children and young delinquents. (Burt, 1925). It was this latter book which gave rise to his nickname of the old delinquent.

At the time Burt was building up educational psychology in London in Newcastle one of Britain's most distinguished psychometrists was devising tests for the selection of children to Grammar Schools. This was Godfrey Thomson, a miner's son from a small Northumberland village who became Professor of Education first at Newcastle and later at Edinburgh where he was head of the Teacher-training College, Moray House. Thomson was the first boy in his village ever to go to Grammar School. He was intellectually brilliant but he knew that many other highly intelligent but poor children were denied a good secondary education. This was because the few scholarships to grammar school depended upon attainment and children from poor backgrounds had no chance. He devised, therefore, the Northumberland Intelligence tests. Intelligence tests are far more just, as selection procedures, to children from poor backgrounds than tests of attainment or interviews. These were, indeed, successful. Poor children went to Grammar Schools and they did well.

At Moray House Thomson developed a huge range of psychometric tests for selection, not only intelligence tests but valid and reliable measures of verbal and numerical ability. Many of these after the Butler Education Act in 1944 were used in England in the now notorious 11+ selection system. The denigration of this selection system is a superb example of the irredeemable irrationality of human beings. Vernon (1961) showed that it was about the most effective selection system which could be devised in terms of its ability to predict school achievement. He also showed that it was far more blind to social class than other selection systems. Yet Pedley (1955) in his powerful advocacy of the Comprehensive School, without recourse to evidence referred to IQ as mystic numbers worshipped by psychologists, scores which condemned poor children to a second-class education.

With the rise of the comprehensive school and the myth that psychometric testing was unfair psychological testing in schools declined. Not only was there no selection but educational psychologists used them far less in their work. This is still true today. Although there is considerable emphasis on testing to ensure that the standards set by the national curriculum are being met, none of these are psychometric tests in the sense of being measures with clear statements of validity and reliability and of a known factorial structure.

The USA is the country where psychological testing has really become established. Semeonoff (1981) recalled how on a visit to the USA even the taxi-driver knew what the Rorschach was. Of course in America generalisations are difficult to make with state laws being different and a considerable reluctance to employ federal legislation. Very large numbers of psychological tests are applied and used daily in America. There is money in testing and private educational psychologists do good business assessing the children of middle-class America. The widespread university education combined with the popularity of Psych 1 in the their courses means that Americans are acquainted with psychological tests and are conditioned to respect their expertise. One possible consequence of all this is the widespread use of quality psychometric tests of attainment and ability for university selection, particularly at the graduate level. The best university graduate schools demand that their applicants sit the Graduate Record Exam and set fixed criteria for entrance. At the undergraduate level Scholastic Aptitude tests are often used. Many of these are produced by the Educational Testing Service at Princeton, which not only constructs these excellent tests of the highest psychometric quality but carries out fundamental research into psychometric methods and has had at it head some of the best psychometrists. The National Foundation for Educational Research in this country it was hoped would play a similar role but it has never received the encouragement from schools and universities that would have won it the necessary funds. However it has devised some good tests and it constitutes a bank of psychometric knowledge in the field.

I shall now turn to the role of psychometric tests in occupational psychology. The assumption behind the use of tests in this field is that there exits for each job an ideal set of personal characteristics and that if the right people are selected for jobs then they are happier and the jobs are better done. This is the Panglossian view of occupational psychology which the purveyors of tests and occupational psychologists strive to maintain.

The first success of psychometric tests in selection came in America when men were drafted into the Armed Forces in World War 1. In all wars there is the problem that men have to be trained in as short a time as possible to carry out jobs of which they have had no previous experience. All this work made it clear that Intelligence tests were valuable in selection at all levels. To jump 50 years Ghiselli (1966) in a study of aptitude tests summarised the results of 10,000 studies. He showed that on average, regardless of job, intelligence correlated .3 with success at that job. No other variable approached this figure. This success of psychometric tests was repeated both the in the USA and in Great Britain in the Second World War. Vernon and Parry (1949) wrote up the British findings in a classic study in applied psychology. Here, inter alia, the interview was shown to be worthless in selection.

It should be obvious that there is more to job success than ability. With the development of good personality tests, based upon the research into the Structure of Personality, which I have already discussed, these tests are now widely used. Cattell's 16 Personality Factor Test is particularly popular as are a host of tests which depend more on clever advertising and selling than psychometric quality. These I must not mention for fear of slander. Serious occupational psychologists have attempted to build up a library of occupational profiles on tests to aid selection and these have proved useful (Cattell and Johnson, 1986). Suffice it to say that today for many graduate jobs both here and in America, and the term graduate is no longer a small minority, applicants will be faced by psychometric tests of ability and personality. I shall conclude this section with a personal observation. Highflying Merchant Banks use intelligence tests for selection of economists because they work better than anything else, knowledge of economics, degree class or any other indicator. For 90 years of psychometrics this is both comforting, there is a general ability, and depressing but what else has been discovered that really works?

Before concluding I want to mention and indicate the nature of some of the technical developments in psychometrics. In general it may be said that many of them are attributable to the enormous improvements in computing. How factor analyses were done before computers is a source of amazement to me given the complexity of hand calculating even a small example. Nevertheless the pioneers of factor analysis were forced to use short-cut methods. Only relatively recently has it been possible to use the most accurate factor analytic procedures on large data sets. Now it is and maximum likelihood procedures offer a statistical basis for the correct number of factors. In addition, Joreskog and Sorbom (1984) have developed a computing procedure, LISREL, which enables users not only to test hypotheses with factor analysis but to utilise even more complex models to understand the relationships between variables. This is known as structural equation modelling. These techniques are certainly running alongside, if not replacing, standard factor analyses.

The classical model of a psychometric test assumed that there was an underlying factor accounting for the correlations between tests. For this reason factor analysis was the preferred method of test construction. Now, however, there are other models of item-response, described in item-response theory,

(Lord, 1980) in which the parameters of responding to individual items are taken into account in test construction. The main use of these models is in the development of item banks and sets of items which are exactly equivalent so that subjects can be retested on different items to check progress or tested once with a brief set of items. Such tests are usually presented on computer where rather than an individual complete the whole test, including items which are much too easy and much too difficult, only items at the threshold level are given. In this way a brief test can quickly yield an accurate score. This technique is known as tailored testing.

The use of computers in testing has also created two more new developments. The first is important in the applied field. The computer can easily be programmed to present the test, score it as the candidate completes it and almost immediately print out the score and its interpretation. This ability to give immediate feedback is valuable, especially where tests are used in appraisal rather than selection or in a medical setting.

The second advantage of the computer test lies in the fact that it is possible to devise items which would be impossible with a pencil and paper test. Items requiring subjects to follow moving stimuli and to respond at speed, time in milliseconds, are obvious examples. Test constructors have not been slow to utilise these advantages which computers offer. However one should not be seduced. A test is no better than its items. Computer tests can be just as bad as any others.

Psychometrics, like this Institute, has ninety years of history. To compress it adequately into forty minutes is not possible, at least for this speaker. I have tried to point out how the issues in the subject have developed over these years, and have concentrated on Great Britain since many of these ideas have arisen here, often from University College. If I were to summarise I would have to conclude that, overall, psychometrics has disappointed. The first basic ideas, factor analysis and the elucidation of structure were good, yet it has not much developed except in technical sophistication beyond that point. The reasons for this are complex but I believe that they include a very basic human fear of being judged inferior. The better psychometrics becomes the greater the fear and so science, for scientists are human, turns against it. That is partly why it has never formed a mainstream part of any department of psychology or education. It is on the fringe. Perhaps also it explains the vilification which many of its leading exponents have suffered. Perhaps, as was long ago suggested, there is some knowledge it is better not to have. I can only say that I do not believe this. To quote Plato, the unexamined life is not worth living. Ignorance is not bliss.

References:

Adorno, T.W., Frenkel-Brunswick, E., Levinson, J. and Sandford, R.N. (1950). *The Authoritarian Personality*. New York, Harper and Row.

Binet, A. (1903). *L'Etude Experimentale de L'Intelligence*. Paris, Sleicher Freres.

Burt, C.S. (1925). *The Young Delinquent*. London, University of London Press.

Burt, C.S. (1955). The evidence for the concept of intelligence. *British Journal of Educational Psychology*, 25, 158-177.

Burt, C.S. (1966). The genetic determination of differences in intelligence: a study of monozygotic twins reared together and apart. *British Journal of Psychology*, 57, 137-153.

Carroll, J. B. (1993). *Human Cognitive Abilities*. Cambridge, Cambridge University Press.

Cattell, R. B. and Johnson, R.C. (Eds) (1986). *Functional Psychological Testing*. New York, Brunner Mazel.

Cattell, R.B. (1957). *Personality and Motivation Structure and Measurement*. Yonkers, World Book Company.

Cattell, R.B. (1971). *Abilities Their Structure Growth and Action*. Boston, Houghton-Mifflin.

Cattell, R.B. (1978). *The Scientific Use of Factor Analysis*. New York, Plenum.

Cattell, R.B. (1981). *Personality and Learning Theory*. New York, Springer.

Cattell, R.B. (1982). *The Inheritance of Personality and Ability.* London, Academic Press.

Eysenck, H.J. (1967). *The Biological Basis of Personality.* Springfield, C.C. Thomas.

Eysenck, H.J. (Ed) (1981). *A Model for Personality.* Heidelberg, Springer-Verlag.

Eysenck, H.J. and Eysenck, S.B.G. (1975). *The Eysenck Personality Questionnaire.* Sevenoaks, Hodder and Stoughton.

Ghiselli, E.E. (1966). *The Validity of Occupational Aptitude Tests.* New York, Wiley.

Gould, S.J. (1981). *The Mismeasure of Man.* New York, W.W. Norton.

Guilford, J.P. (1959). *Personality.* New York, McGraw-Hill.

Guilford, J.P. (1969). *Human Intelligence,* New York, McGraw-Hill.

Hearnshaw, L.S. (1979). *Cyril Burt: Psychologist.* London, Hodder and Stoughton.

Horn, J. and Knapp, J.R. (1973). On the subjective character of the empirical base of Guilford's Structure of Intellect Model. *Psychological Bulletin,* 80, 33-43.

Hudson, L. (1966). *Contrary Imaginations.* Methuen, London.

Joreskog, K.G. and Sorbom, D. (1984). *LISREL.* Chicago, International Educational Services.

Lord, F.M. (1974). *Applications of Item Response Theory to Practical Testing Problems.* Hillsdale, Erlbaum.

Kamin, L.J. (1974). *The Science and Politics of IQ.* Harmondsworth, Penguin.

Kline, P. (1993). *Personality: The Psychometric View.* London, Routledge.

Kline, P. (1997). The relation between personality and ability. Paper at The Spearman Conference, July, 1997, University of Plymouth.

McClelland, D.C. (1961). *Achieving Society.* Princeton, Van Nostrand.

McCrae, R.R. and Costa, J.P. (1990). *Personality in Adulthood.* New York, Guildford Press.

Pedley, R.R. (1955). *The Comprehensive School.* Harmondsworth, Penguin.

Semeonoff, B. (1981). Projective Tests, In Fransella, B. (Ed.) (1981). *Personality.* London, Methuen.

Spearman, C. (1904). General intelligence, objectively determined and measured. *American Journal of Psychology,* 15, 201-293.

Terman, L.M. and Oden, M. (1959). *The Gifted-Group at Mid-Life*. Stanford, California UniversityPress.

Thurstone, L.L. (1947). *Multiple Factor Analysis. A Development and Expansion of Vectors of the Mind*. Chicago, University of Chicago Press.

Vernon, P.E. (1961). *The Measurement of Abilities*. London, University of London Press.

Vernon, P.E. (1979). *Intelligence, Heredity and Environment*. New York, W.H. Freeman.

Vernon, P.E. and Parry, J. (1949). *Personnel Selection in the British Forces*. London, University of London Press.

The Galton Lecture 1997: The Eugenics Society and the Development of Biometry

A W F Edwards

'Biometry' was the title of the opening article which Francis Galton contributed to the new journal which he, W F R Weldon and Karl Pearson founded in 1901. No-one looking at *Biometrika* today would call it, as did they, 'A Journal for the Statistical Study of Biological Problems', a heading which was quietly dropped just fifty years ago, at the very moment, 1947, when the *Biometrics Bulletin* of the American Statistical Association was transmuting itself into *Biometrics*, the journal of the newly-founded International Biometric Society.

As is well-known, Pearson never joined the Eugenics Education Society which his hero Galton had been instrumental in starting, and whose ninetieth birthday we celebrate at this meeting, but with the abandonment of *Biometrika* to the mathematicians the founding of *Biometrics* ensured that the subject biometry was to retain its link with Galton and the Eugenics Society by another route. For my own teacher, R A Fisher, was not only linked to Galton by the many ties I described in my lecture 'Galton, Karl Pearson and modern statistical theory' at the 1991 Symposium, but he was the first President of the International Biometric Society. At the inaugural meeting of the British Region of the Society in 1948 he gave an address in which he claimed that it was biometry, and not surveying or astronomy, which had taken the step of making known the principles of induction. He said 'But, as it happened, it has been reserved for Biometry, the active pursuit of biological knowledge by quantitative methods, to take this

DEVELOPMENT OF BIOMETRICS

great step; and the man who in the nineteenth century did more than any other to prepare the way was, I think, undoubtedly Francis Galton'.

The word 'biometry' of course keeps gently evolving, and Galton himself was responsible for replacing William Whewell's original 1831 meaning 'calculations on lives' with 'the application to biology of the modern methods of statistics'. Mrs Weldon, in bequeathing money to University College London for a Professorship of Biometry, defined it as 'the higher statistical study of Biological problems', a remit met with distinction by the first two professors, J B S Haldane and C A B Smith. Largely because of the historical symbiosis of biological statistics and population genetics we nowadays think of biometry as encompassing more of biological mathematics than just the statistical, and the International Biometric Society describes itself as being 'devoted to the mathematical and statistical aspects of biology'. But I prefer Fisher's definition 'the active pursuit of biological knowledge by quantitative methods' because it stresses that it is biological rather than mathematical knowledge that biometry seeks. It is in this sense that I shall trace the contributions that the members of the Eugenics Society made, principally through the columns of the Society's journal *The Eugenics Review*, to the development of biometry.

I am particularly grateful to the Council of the Society, or, as I must now call it, the Galton Institute, for inviting me to give the 1997 Galton Lecture, because it gives me an opportunity to repay a debt to the Society incurred at the beginning of my career when I held its Darwin Research Fellowship in 1960-61. I supposed at the time, and for the ensuing thirty years, that the Darwin was Charles, but I was delighted to discover from J H Bennett's book *Natural Selection, Heredity and Eugenics* that it was in fact Charles' fourth son Leonard, the long-serving President of the Society. Even the article 'The activities of the Eugenics Society' in the closing volume of *The Eugenics Review* in 1968, by F Schenk and A S Parkes (**60**, 142-161), failed to record that the Darwin of the Research Fellowships was Leonard, though perhaps the authors thought everyone knew.

The last time I met Fisher was during my tenure of the Fellowship, but he never told me its history, and it was not until I had had an opportunity to examine his correspondence now lodged at the University of Adelaide that I learnt that the creation and naming of the Fellowships was his idea. He wrote 'I think we should all like to pay a tribute to the great work which Major Darwin has in the past done for the Society by associating his name with the proposed Studentships as by designating them the Leonard Darwin Studentships in Eugenics'. He also referred to their creation, I now see, in his address to the Annual General Meeting of the Society in May 1935, printed under the title 'Eugenics, academical and practical' in the *Review* later that year (**27**, 95-100). I do not know what happened to them, but I was very grateful for mine, and perhaps the Institute should start them again.

I hope that one day someone will produce a variorum edition of Fisher's *Statistical Methods for Research Workers*, whose fourteen editions ran from 1925 until a posthumous one, following his death in 1962. Should they do so, the test of their attention to detail will be the entry 'Geissler' in the list of references, which remained without an initial until the thirteenth edition, when it acquired an 'A', for Arthur. You may think it absurd that I should know this, and, knowing it, that I should tell you. I know it only because it was I who informed Fisher that Geissler's Christian name was Arthur, as a result of which he made a note in the new edition of *Statistical Methods* he was then preparing. The reason why I am telling you will have to wait, for it is part of a detective story.

Probably the most celebrated argument in quantitative biology is the explanation of the near-equality in the numbers of males and females in many species which R A Fisher gave in

The Genetical Theory of Natural Selection in 1930. It demonstrated how natural selection operating at the individual level could mould even a population characteristic such as the sex-ratio which had been considered an obvious candidate for between-population selection; it showed how sometimes it was necessary to consider three generations and not just two in an evolutionary model; it was hailed as the first example of an ESS (evolutionarily stable strategy) by those who later christened the concept; and it has frequently been clothed in game-theoretic language as a key example by those to whom such an approach appeals. It even started the modern interest in the evolutionary implications of parental expenditure. It greatly influenced W D Hamilton's 1967 paper 'Extraordinary sex ratios' and the views of G C Williams. Hamilton's papers and Williams's 1966 book *Adaptation and Natural Selection* have been held to be mainly responsible for the triumph of the view that natural selection operating within populations is the primary mechanism of evolution, so superbly argued by Richard Dawkins in his series of books. Both the sex-ratio argument and the concept of parental expenditure were attributed to Fisher by Dawkins in *The Selfish Gene* in 1976.

I wonder if the name J A Cobb means anything to you. It will if you are familiar with the later chapters of *The Genetical Theory of Natural Selection* because Fisher there refers to Cobb's 'brief but important note' 'Human fertility' in *The Eugenics Review* for 1913 (4, 379-382) as the source of the theory of the social selection of human fertility. Cobb had developed this theory out of Galton's explanation for the relative infertility of heiresses which he had expounded in *Hereditary Genius* in 1869, and Fisher viewed it as the major cause for eugenic concern affecting the British population between the wars. Until recently I had never seen anything else of Cobb's apart from this note.

It was therefore with astonishment that, on choosing the 1914 volume of *The Eugenics Review* for my initial foray into the literature for the purposes of this lecture, I happened to open it at 'The problem of the sex ratio' by J A Cobb (**6**, 157-163). This had been my thesis subject. I was supposed to have been breeding mice, but I had become fascinated by Fisher's account of the binomial distribution in *Statistical Methods*, for which he had used Geissler's data on the sex ratio, and in particular Geissler's figures for the numbers of boys and girls in families of eight children. Moreover, my growing interest in the subject had led me to ask Fisher the obvious question about natural selection and the sex ratio: 'Why is it that in herd animals where dominant males monopolize the mating, natural selection has not adjusted the sex-ratio at birth accordingly?', and Fisher had made the obvious answer: 'Go and read *The Genetical Theory*', which I did. Yet neither there, nor anywhere else, have I ever seen Cobb's paper referred to.

It is a truly remarkable piece. It not only contains what we all call 'Fisher's' argument on natural selection and the sex ratio, complete with a discussion of the influence of parental expenditure, but also some very advanced statistical arguments concerning the evidence for heritable variation in the sex ratio, in which Cobb uses Geissler's data for families of eight children. In his own words,

> If we take the sex-ratio at birth it appears at first sight that the numbers of the sexes born will become equal. For if there are more born of one sex, say, the male, a female will have a greater chance of finding a mate than a male. There will be more matings, therefore, among the descendants of mothers of females than amongst the descendants of mothers of males. The mothers of females will therefore be better represented in the third generation, and as their characteristic is assumed to be inherited, there will be a tendency for the sex-ratio to diminish until it reaches equality in numbers between the sexes at birth. The same argument will show a tendency towards equality between the numbers of the sexes at the time of conception and at the age of marriage.

> The numerical equality of the sexes may therefore be accounted for in a general way on the ground of heredity, not as has been often said, because as between race and race that race would survive in which the sexes were nearest equality, but because, as I maintain, within the race those individuals who tend to reduce the inequality of the sex-ratio will have more descendants.

Cobb goes on to discuss the rôle of parental expenditure exactly as Fisher was later to do in *The Genetical Theory*, but it would take me too far out of my way to describe his contribution, and its relation to earlier work by Carl Düsing and Corrado Gini, further today, and I must reluctantly limit myself to providing the evidence that Fisher was familiar with Cobb's paper. The similarity of treatment and the fact that Fisher wrote in the same number of *The Eugenics Review* is almost evidence enough, but when we see that Fisher's use of Geissler's data was limited to the families of size eight which Cobb had also used, no doubts remain. Fisher had needed an example of the binomial distribution for his textbook, and here was one to hand. Why look further? Why, indeed, bother to seek out the original? So he took the reference to Geissler from Cobb's footnote, and, give or take an umlaut or two, it is identical, *including the absence of A for Arthur.*

In the full account of this story, which will be published elsewhere, I trace the ideas about natural selection and the sex ratio back to Charles Darwin in the first edition (but not the second) of *The Descent of Man and Selection in Relation to Sex* (1871), and I explain that Fisher's lack of a reference to the paper of Cobb's, and Cobb's lack of a reference to anyone on that topic, probably mean no more than that the ideas were circulating freely at the time amongst the small number of people involved, who regarded them as being in the public domain. It is not they who have been ignorant, but us. After all, you cannot even learn from *The Genetical Theory* that the famous argument for stable equilibrium due to heterozygotic advantage was published by Fisher himself eight years earlier.

I have been unable to find out anything further about J A Cobb, whose name is not to be found in any of the three excellent books on which I rely so much, Joan Fisher Box's life of her father *R A Fisher, The Life of a Scientist*, Bennett's *Natural Selection, Heredity and Eugenics*, and Pauline Mazumdar's *Eugenics, Human Genetics and Human Failings*.

Here, then, is my first example of a quantitative biological argument to emerge from the Eugenics Society, and it turns out to be one of the most influential of the century. But now it is time to pass on to a more systematic examination of *The Eugenics Review* and see what else we can find. Astonishingly, moving on from the evolution of the sex ratio and the influence of parental expenditure, we next find ourselves face to face with the founding concept of the population genetics of altruism, a third idea of great influence on modern evolutionary biology through Bill Hamilton's mathematisation of it with 'inclusive fitness' in 1964. Inclusive fitness gives exact recognition to the fact that helping one's genes to survive by caring for one's children is only a special case of helping one's genes by caring for any other blood-relative, for there is a calculable chance, depending on the degree of relationship, that he will share each of your genes through your common ancestry. Hence any gene for such altruistic behaviour may flourish.

The Eugenics Review for 1914 contains the key (**5**, 309-319), at the end of a paper 'Some hopes of a eugenist' by R A Fisher which he had read to the Society in October 1913 and which, he tells us, is based on his paper to the Cambridge University Eugenic (*sic*) Society in November 1912, when he was a 22-year old postgraduate. Fisher is discussing 'the cooperation of different genetic types' within a single population such as 'bees in a hive'. How can such versatility be maintained? Fisher argues through an example: 'Suppose ... that a group of distinguished families possess ... versatility to the extent of being able successfully to fill the rôle, either of a landed

DEVELOPMENT OF BIOMETRICS 163

gentleman ... or of a soldier. A is the eldest son, and stays at home; his brother B goes to the wars; then so long as A has some eight children, it does not matter, genetically, if B gets killed, or dies childless, there will be nephews to fill his place'.

I fear a little biometrical disaster has taken place in the calculation, in the best Galtonian tradition. Galton, it may be remembered, thought that if judges had on average four siblings, they would expect to have one-and-a-half brothers and two-and-a-half sisters, since in his day all judges were men. He thus made the first, but certainly not the last, ascertainment error. Our 22-year-old postgraduate has, I suggest, made the first error in computing gene identity-by-descent (a concept, incidentally, not supposed to have been invented until 1940). Perhaps he has worked out that the probability of a nephew carrying a particular one of A's genes identical by descent is 1/8, as indeed it is, but has then forgotten that A carries two genes at each locus so that overall the expected number is 1/4, not 1/8, and only four nephews are needed to compensate for B's childlessness. Even if we add the two children which A will need to ensure that his own genes are fairly represented in the next generation we still only reach six children in all. Or perhaps Fisher has thrown in a factor of two because landed gentlemen and soldiers were necessarily male. It does not matter very much, and we should in any case not forget that knowledge of Mendelian inheritance was only twelve years old in 1912.

As was to be expected, the idea resurfaces in *The Genetical Theory*, where Fisher uses it to explain the evolution in insects of distastefulness to predators. Distastefulness, being a quality only in evidence on the death of the insect, has nevertheless evolved because the experience of eating a distasteful larva will persuade the predator not to try nearby larvae, who, sharing genes with the unfortunate victim, thereby confer on his genes for distastefulness a selective advantage. 'The principle', wrote Fisher, is that 'of the selective advantage shared by a group of relatives, owing to the individual qualities of one of the group, who enjoys no personal selective advantage'. From *The Genetical Theory* the idea found its way into the extremely influential work of Hamilton as 'inclusive fitness', and thence into a primary rôle in evolutionary biology. To us it is not only a matter of pride that *The Eugenics Review* carried the idea, but a matter of interest that it was eugenic considerations that prompted it.

I must pass over Fisher's fundamental contribution to Darwin's theory of sexual selection, contained in volume 7 of the *Review* (184-192; 1915) on the grounds that it is insufficiently quantitative to count as biometry, and pass on to a paper, not in the *Review*, whose last sentence is 'Finally, it is a pleasure to acknowledge my indebtedness to Major Leonard Darwin, at whose suggestion this inquiry was first undertaken, and to whose kindness and advice it owes its completion'. I refer, of course, to the foundation paper of biometrical genetics, generally known simply as 'the 1918 paper' in the manner of 'the 1812 overture', Fisher's 'The correlation between relatives on the supposition of Mendelian inheritance' published in the *Transactions of the Royal Society of Edinburgh* (**52**, 399-433). It is a difficult paper, and Fisher wrote a layman's account for the *Review* (**10**, 213-220, 1918) in which he summarised his theoretical findings: '(1) The facts of Biometry do not contradict, but in many ways positively support the theory of cumulative Mendelian factors. (2) If this theory is correct a sufficient knowledge of the correlation coefficients for any one feature, between different pairs of relatives, would enable us to analyse completely and estimate numerically the percentage of variance due to heritable factors'.

There is no need for me to emphasise the central position of this work in securing the foundations of modern biometry. Its influence is felt not only in genetics and animal and plant breeding, but throughout the whole of statistics. The word 'variance' was coined in the main paper, but to *The Eugenics*

Review is reserved the distinction of having printed the phrase 'analysis of variance' for the very first time. Fisher's *Review* paper also ends with thanks to Darwin, but not before its author has written 'In conclusion it is right that I should express my deep sense of gratitude to the Eugenics Education Society, who have most generously assisted me throughout'. This presumably refers in part to the contribution that the Society made to the costs of publication of the main Royal Society of Edinburgh paper, which that Society had required. Nor should we overlook the rôle of Leonard's brother Francis who, in the first ever Galton Lecture, in 1914, (*Eugenics Review*, **6**, 1-17) had stressed the need to reconcile the Biometric and Mendelian approaches. Fisher's developments were well summarised six years later in two lectures at the London School of Economics printed as one article 'The biometrical study of heredity' in the *Review* for 1924 (**16**, 189-210). A small historical point is that Fisher's well-known argument in the opening paragraphs of *The Genetical Theory* about the statistical consequences of blending inheritance occurs for the first time in this article.

Stepping back to 1918, the next contribution from the *Review* of note is somewhat specialised, but of major importance. Fisher computes the chance of survival of a new mutation, the first time stochastic considerations in genetics have been mathematically assessed. In 'Darwinian evolution of mutations', dated February 2nd, 1921 but published in 1922 (**14**, 31-34), he iterates a generating function to estimate the chances of survival. This advance is normally credited to his more technical paper of the same year in the *Proceedings of the Royal Society of Edinburgh* (**42**, 321-341), but that paper is headed 'Read June 19, 1922', so we may claim priority for the *Review*. The method makes another appearance in chapter IV of *The Genetical Theory*.

The *Review* (**19**, 103-108, 1927) may next claim the distinction of having carried Fisher's introduction of the concept of 'reproductive value', an idea peculiarly appropriate for biometry given the original meaning of that word. Taking his cue from Malthus, who had drawn an analogy between population increase and compound interest, Fisher extended the actuarial treatment of mortality to reproduction, thus formalising the notion of the contribution of an individual to the ancestry of future generations. The parameter which he called 'the Malthusian rate of increase' turned out to be the same as the 'intrinsic rate of natural increase' which A J Lotka had introduced in 1925. Since it has been suggested that Fisher borrowed Lotka's idea when he introduced his Malthusian parameter in *The Genetical Theory* five years later, it is encouraging to see it clearly set out as early as 1927, and in fact Fisher acknowledged the similarity after Lotka wrote a letter to the *Review* (**19**, 257-8, 1927) when he stated that he had not previously seen Lotka's work, and added 'Evidently the only absolutely novel suggestion in my article lies in the estimation of a definite "reproductive value" for each age of life'.

We have now travelled through the pages of *The Eugenics Review* from 1914 as far as 1930, picking out the contributions to biometry recorded in them. It is time to summarise our findings. Natural selection and the sex ratio, parental expenditure, inclusive fitness, biometrical genetics, the analysis of variance, the chance of survival of new mutations, the Malthusian parameter of population increase, reproductive value. Perhaps we should not be surprised that all these ideas passed through the mind and the pen of a single writer, R A Fisher. Some of them were new and wholly original to him, such as inclusive fitness, the survival of mutations, and reproductive value. One, the Malthusian parameter, we now know was published independently a couple of years earlier. Another, the treatment of natural selection and the sex ratio, was published in the *Review* by J A Cobb but made famous by Fisher in *The Genetical Theory*. The Society, by assisting in the publication of Fisher's 1918 paper and supporting him in other ways, contributed handsomely to the subsequent development of biometrical genetics, whose origins, as Fisher recognised, could be traced back to Udny Yule and Karl Pearson. The seed-corn for the analysis of variance was planted in the 1918 paper, and if its full growth took place later at Rothamsted, at least the Society can proudly display the coining of the phrase in its own *Review*.

Over all this activity presided the genial, fair-minded, uncontroversial figure of Leonard Darwin, President of the Society from 1911 to 1929, whom Fisher described on his death as 'surely the kindest and wisest man I ever knew'. 1930 is therefore a good place to pause, and additionally so because that was the year in which *The Genetical Theory of Natural Selection* appeared, containing revised versions of nearly all the biometrical arguments which Fisher had previously published in *The Eugenics Review*. How very appropriate, therefore, to recall the dedication with which Fisher opened his famous book:

TO

MAJOR LEONARD DARWIN

*In gratitude for the encouragement,
given to the author, during the last
fifteen years, by discussing many
of the problems dealt with
in this book.*

We should not take our leave of Leonard Darwin without remembering Bernard Darwin's verse about him, quoted by Gwen Raverat in *Period Piece*:

> Serenely kind and humbly wise,
> Whom each may tell the thing that's hidden
> And always ready to advise
> And ne'er to give advice unbidden.

As a matter of fact Leonard's father himself was not immune from clerihews in the columns of *The Eugenics Review*, for to my surprise I discovered during my search for serious material one that I had written myself. In 1965 the editor of the *Review* had lifted from *The Lancet* of August 22nd, 1964

> Darwin is a witness
> To the value of unfitness.
> He survived as a loafer
> On a sofa.

to which I had responded (*Review* **57**, 48; 1965)

> That Darwin's a loafer I'm bound to admit
> But why should this slothfulness make him unfit?
> I thought he himself was the principal witness
> That number of children determines one's fitness.

This ignores inclusive fitness, so I feel obliged now to add

> But Hamilton's taught us, what Fisher first knew,
> To credit the genes of our relatives too,
> Each reckoned by kinship for what he is worth
> And added to those who are counted by birth.

Gwen Raverat, who was one of Charles Darwin's granddaughters, was herself rather clear about this, for Chapter X of *Period Piece* opens 'One year, at the Christmas party, all the five uncles were there; and among uncles I include my father. A father is only a specialized kind of uncle anyhow'.

But we must return to serious matters. From 1930 onwards the scene changes dramatically, as Mazumdar has so well chronicled in her book. Politically the climate altered as the left, in the persons of Lancelot Hogben and J B S Haldane, came to challenge the initial enthusiasms of the revolution wrought by the realisation that Charles Darwin's theory applied to man. (1930 is closer to the publication of *Hereditary Genius* and *The Descent of Man* than it is to us.) Biometrically, so to speak, the climate changed too, as the statistical advances of the German school of eugenists were made known in Great Britain, principally by Hogben. In his *Nature and Nurture* lectures given in the University of Birmingham in 1933 Hogben criticised the splitting of the total variance into genetic and environmental components contained in Fisher's 1918 paper (Hogben's book, incidentally, is dedicated to Haldane), rather as Leonard Darwin had so presciently managed to do in 1913. Hogben's lectures ended with a quotation from Wilhelm Ostwald, clearly directed at Fisher:

> Among scientific articles there are to be found not a few wherein the logic and mathematics are faultless but which are for all that worthless, because the assumptions and hypotheses upon which the faultless logic and mathematics rest do not correspond to actuality.

Fisher was quite relaxed in his response; in a letter to Hogben he said:

> I think I see your point now. You are on the question of non-linear interaction of environment and heredity. The analysis of variance and covariance is only a quadratic analysis and as such only considers additive effects. ... perhaps the main point is that you are under no obligation to analyse variance into parts if it does not come apart easily, and its unwillingness to do so naturally indicates that one's line of approach is not very fruitful.

The argument rumbles on to this day, and I myself am inclined to side with the sceptics. People become excited about the heritability of intelligence, for example, but there can be no such thing in the absolute. All there can ever be is the computation of the heritability of intelligence for a specific set of data, based on the unrealistic linear model. But what does it tell you? Only that if the environment had been more uniform the heritability would have been higher, which you knew already.

But it is characteristic of these years that Fisher, Haldane and Hogben managed to preserve quite good working relations most of the time in spite of their political differences. All three were enthusiastic proponents of constructing a linkage map for man using the blood-group genes as markers, and Hogben persuaded the Medical Research Council to set up a Committee for Human Genetics with Haldane as Chairman and Fisher amongst its members, who also included L S Penrose and J A Fraser Roberts. An important result of this upsurge in interest in human genetics in London was the support that the Rockefeller Foundation gave to Fisher to set up a serology unit after he became Galton Professor of Eugenics at University College in 1933. In a lecture at the 1935 Annual General Meeting of the Eugenics Society to which I have already referred, Fisher outlined his vision for the unit, that discovery of the linkage relations amongst disease genes and normal marker genes such as the blood groups would revolutionise prognosis by enabling deleterious recessives or late-onset dominants to be detected in normal individuals, knowledge that could then be used in genetic counselling. The human genome project was on its way.

I am in danger of departing too far from my biometrical remit, but I would just like to quote what Hogben had to say about Fisher's views on the social selection of fertility contained in *The Genetical Theory of Natural Selection* which I have already briefly mentioned in connection with J A Cobb, for it might have been supposed that Hogben would have been hostile. In *Genetic Principles in Medicine and Social Science* (1931) he wrote:

> R A Fisher has pointed out that low fertility assists materially in social advancement in societies in which mercantile or industrial interests predominate and thrift is the supreme social merit. This might tend to concentrate individuals who are incapable of having large families in the governing classes. As a long view of the situation, Fisher's argument is worthy of attention. For the collective endowment of parenthood

he makes a case which compels the serious consideration of those who are disposed to regard private enterprise in family production as an institution sanctified by natural law.

But I digress too far. The years after 1930 belong to the development of linkage and other statistical methods specifically directed at man. Haldane and Fisher took the baton from Hogben and from 1934 onwards developed the methods which led ultimately to today's computer programs. But Hogben never let up on his attacks on modern statistical methods in general, and Fisher's in particular, culminating in his massive and extremely well informed *Statistical Theory* published in 1957 with the description 'An examination of the contemporary crisis in statistical theory from a behaviourist viewpoint'. It sank without trace. When, a dozen years later, I was reading everything I could lay my hands on whilst writing my own book on statistical inference *Likelihood*, I did not come across it. Nor, evidently, did Ian Hacking in his influential *Logic of Statistical Inference* (1965).

The Eugenics Society was not a direct force in these technical advances in linkage theory, but it did contribute to them nevertheless. Together with the Galton Professorship Fisher had acquired the editorship of the *Annals of Eugenics* which carried much of his and Haldane's linkage work, and which, as the *Annals of Human Genetics* since 1954, continues to report on linkage to this day. For as Professor J S Jones reminded us at the 1991 Symposium, the *Annals* was jointly published by the Galton Laboratory and the Eugenics Society from 1934 until 1940, the support of the Society enabling Fisher to institute quarterly publication.

Looking down the list of Galton Lecturers since 1914 it is not surprising that we do not find the name of Hogben, or of Haldane for that matter. But it is surprising not to find Fisher. Perhaps he felt that he had said enough in his 1932 Herbert

Spencer lecture in Oxford, *The Social Selection of Human Fertility*, which was his main eugenic concern, or that his 1935 'Eugenics academic and practical' was *de facto* a Galton Lecture. When Fisher's secretary opened the parcel containing his pre-publication copies of the Dover edition of *The Genetical Theory of Natural Selection* in 1958 I happened to be standing nearby. 'Hmm', he said, observing the design on the front of the paperback, 'it looks like a book by Hogben'. But I have an even better Hogben–Fisher story to leave you with. Neither of them may have been Galton Lecturers, but it involves two people who have been, my brother Professor J H Edwards, the 1991 Lecturer, who was influenced by Hogben, and me, a student of Fisher's, and I record it for posterity with my brother's permission. It is from a letter from Fisher to R R Race, the serologist, written in 1960:

> The Edwards who messed up Cleghorn's data, and is formally thanked for it in his letter to *Nature* seems not to be my Edwards from Cambridge. It was the thought that it was he that annoyed me, for the estimates published in *Nature* were manifestly incompetent, and I feared that one of my own pupils was running amok and adding unnecessarily to darkness and confusion. However, I understand he is only one of Hogben's, so all is explained.

I like to think that Galton would have been amused. I am sure he would at least have been delighted to discover another pair of brothers amongst his Lecturers, and for the opportunity to complete the second pair I am deeply grateful.

Eugenics in France and in Scandinavia: Two Case Studies

Alain Drouard

Introduction

I am grateful to the Galton Institute for this opportunity to present to an English readership this paper on aspects of eugenics in Europe.

As a historian I have devoted myself to the study of these questions over many years, beginning with my thesis on the French Foundation for the Study of Human Problems created by one of the most important figures of eugenics in France the Nobel Prize winner for medicine Alexis Carrel (1873-1944), also the author of a widely acclaimed best seller *Man the Unknown* published in English and in French in 1935.[1]

The importance of eugenics is as clear today as in 1935. There can be no understanding of the issues raised in our societies, especially by the development of new techniques of procreation, without references to eugenics, a word coined by Sir Francis Galton to define:

> "the science of improving stock, which is by no means confined to questions of judicious mating, but which, especially in the case of man, takes cognisance of all influences that tend in however remote a degree to give to the more suitable races or strains of blood a better chance of prevailing speedily over the less suitable than they otherwise would have had."[2]

After Nazism and the crimes perfected in its name (mass extermination, euthanasia), eugenics has become a taboo.[3] But we have to remind ourselves that the history of eugenics does not lie in this criminalisation process. Since its origins eugenics has been diverse and influential in theory and in practice.[4]

In this paper I will deal mainly with eugenics in France and secondly but more briefly with eugenics in Scandinavia, which together constitute two case studies of eugenics in a democratic context. In France one can observe a large gap between theory and practice where a paradox is now prevailing: although it is constantly denounced it is nevertheless practised every day. The Northern States of Europe - Denmark, Sweden, Norway, Finland have experienced between the two World Wars what can be called a "democratic" eugenics as part of the process of construction of the Welfare State inasmuch as it coincides with a policy of social reforms conducted by the social democrats. On the one hand eugenics had almost disappeared in the Northern States during the seventies and these countries seem to have forgotten this chapter of their history -the recent controversy over Swedish sterilisations illustrates this collective amnesia- but on the other hand they are pioneers in promoting public debates on bioethical issues.

Although the comparative approach is useful and necessary to characterise the differences between the eugenic movements - French eugenics is different from the English which in turn is different from the American - it does not furnish a complete understanding of eugenics.

When considering the elements which are invariably present, one can define eugenics as a form of "scientific ideology"[5] that is to say a doctrine which combines heterogeneous proposals and elements: some inspired by the science of heredity or relating to it and others arbitrary concerning the "fits" and the "unfits", the "superior" and the "inferior", the "feeble minded", the "degenerate people". As an ideology eugenics aims at a global explanation of modern society, of its degeneracy and its possible regeneration. Finally it overlaps with other doctrines such as social Darwinism, racism and social hygiene but without ever being absorbed by them.

It is worth noting that in the case of eugenics ideology has preceded science (genetics). What was called at the end of the nineteenth century heredity is now genetics and what is at stake is the question of its limits. But what must be stressed is the fact that the progress of genetics did not mean the end of ideology. In the first decades of this century genetics undermined the basic assumptions of eugenics while demonstrating its inefficiency, especially the postulate that it could be possible to eliminate heredity diseases by forbidding the defective people to get married and to procreate. The rediscovery of Mendel's laws proved the recurrent character of defective genes - i.e. they appear only when the two parents are bearing them and when the child has received both copies. As a result an efficient eugenic policy should eliminate the degenerate ones but also the bearers of defective genes. And the bearers are more numerous than the sick themselves! For example if one out of 10,000 is defective, it means that one out of 50 is a bearer. The elimination of the defective people could not alone influence the frequency of the defective genes.

Eugenics has been the dominant ideology of the biologists and especially the geneticists during the thirties although the most famous of them denounced racism and Nazism. J B S Haldane condemned racism and Nazism and in 1936 Julian Huxley denounced the dangers of a so-called science represented by the racial Nazi theory. In 1935 the Nobel Prize winner Hermann S Muller denounced the corrupted eugenics of professionalised eugenicists who are the advocates of class and racial prejudices.[6]

The Manifesto of Geneticists adopted at the 7th International Congress of Edinburgh in 1939 while condemning the doctrine of preservation of the race defended eugenics thus:

"The genetical features of each generation could become superiors to those who preceded only through selection so that those who possess a better genetical equipment reproduce more than the others either by a free choice or as a consequence of their behaviour"

Eugenics in France

In France the word eugenics - though first used by Georges Vacher de Lapouge in 1886[7] is found in combination with other words until 1914. Let us quote some of them: "good birth", "conscious procreation", "puériculture", "hominiculture", "human selection", "eugénnétique", "viriculture".

In France doctors and physicians were the first propagandists of eugenic ideas many years before Galton coined the term and the creation of the French Eugenic Society in 1913 and this medical interest is a characteristic feature of French eugenics.[8] At the end of the nineteenth century eugenics was defended by the anarchist and revolutionary neo-Malthusians of the League of Human Regeneration founded in 1896 by Paul Robin and by the pro-natality movement called "Alliance nationale contre la dépopulation" also founded in 1896.

Two Nobel Prize winners for medicine - Alexis Carrel (1912) and Charles Richet (1913) favoured eugenics though their respective approaches were different. Whereas Richet urged the elimination of degenerate children at their birth through euthanasia in *The Human Selection* (1913)[9], forbade marriage of degenerate and abnormal people, recommended castration and sterilisation of criminals, was a neo-Malthusian when considering world population and encouraged natality only in France, Carrel defined a more moderate eugenics, natality oriented, "positive" and "voluntary" in his best seller *Man the Unknown* first published in 1935 both in French and in English. World famous biologist Jean Rostand must also be mentioned as a significant figure in French eugenics.[10]

The diversity of origins and of orientations does not exclude convergence between the different trends of eugenics whose history has not been written so far.[11]

Before Georges Vacher de Lapouge introduced the word in France in 1886, a great number of doctors tried to fight against degeneracy by suggesting the selection of those who wish to procreate as the best remedy. In the mid-nineteenth century French doctor Prosper Lucas defined the criteria of the choice of mates and the best circumstances for procreation. In his view those who were ill should not reproduce and what was a moral constraint should become a legal obligation. He wanted to introduce a compulsory medical examination before marriage.[12] Doctors were coping with a contradiction between their attachment to the principle of secrecy and their wish to see the State regulate procreation with their help. In 1865 another French doctor Alfred Charles Caron coined the term "puériculture" (childcare and feeding)[13] which Adolphe Pinard praised at the beginning of the twentieth century as a major contribution to pro-natality eugenics.

Count Georges Vacher de Lapouge-the first theoretician of eugenics in France-was born in 1854 near Poitiers into family of the old nobility.[14] Although he studied law and medicine he was fundamentally a naturalist and is well known for being an entomologist (he specialised in the Carabini). When preparing the "agrégation" (the highest law degree which leads to an academic position), he discovered Galton, Haeckel and the French school of anthropology of Broca. He met Topinard who supported his first researches. But he never managed to become a University Professor and remained consequently isolated. He mentioned the following reasons in a letter sent to Herman Lundborg in 1927:

"In France the politicians have assassinated anthropology as in Italy the theologians have tried to assassinate astronomy." And he explained: "Around 1885 Broca was dead but anthropology was well alive without many friends but without enemies. Topinard was in charge and I was his faithful ally. By chance an open-minded man was appointed to the Direction of Higher Education, M Liard who had been my philosophy teacher at the high school of Poitiers and my personal friend. I had just elaborated my theories of social selection. Out of interest in the new science and maybe for myself he decided to appoint me as a librarian at the University of Montpellier and I was in charge of a lecture on anthropology at the Faculty of Sciences which he intended to turn into a chair within two years. The lecture was so successful that it attracted students from all over the world and students with scholarships were sent from the United States to attend the lecture on the uses of anthropology in social sciences which I created at the Faculty of Humanities in 1886. It seemed to be the beginning of a new era. Then difficulties began." And Lapouge goes on with his story to conclude: "...The dogma of equality of men is the doctrine of the State and all opposite views are forbidden".[15]

He was appointed a librarian at the University of Montpellier in the South of France and later in Rennes. He lectured with success on anthropology in Montpellier again attracting many students from abroad. These lectures were first published under the title of "Leçons de Montpellier" in the *Revue d'anthropologie* before becoming books: *The Social Selections* (1896), *The Aryan, his Social Role* (1899) and *Race and Social Environment* (1909).

His eugenics is inspired by Galton. Lapouge is almost- it must be stressed - the unique representative of Darwinist eugenics in France.[16] His eugenics is hereditarian, racist and *socialist!* Lapouge was a member of the French Workers Party - a Marxist and collectivist party founded by Jules Guesde in 1893.

He advocated racial selection as the only remedy for the exhausted old democracy:

> "Socialism is more and more the true opposition to the plutocratic democratic regime ... Socialism will be for selection; if not it will not exist".[17]

At the end of his life Vacher de Lapouge hoped that the United States would be able to promote the program of regeneration of the human race that he had conceived: "The achievement of artificial selection is only a question of time. It will be possible totally to renew humanity in a couple of centuries and to change the mass by a mass more superior in which the process of improvement could go on ... Americans, the salvation of civilisation is depending upon you as the emergence from your people of half gods".[18]

Even though he remained isolated in France, Vacher de Lapouge influenced the eugenic movement in Europe and in the United States: his work was translated into several different languages, notably German which led to an abundant correspondence with German eugenicists such as Otto Ammon, Hans F K Günther and Lundborg in Sweden who admired him. He died in 1936.

A second trend in French eugenics was represented by the "puériculteurs" who followed Adolphe Pinard.[19]

Adolphe Pinard, the son of peasants from Champagne, was born in 1844. He had a brilliant career as a doctor, specialising in obstetrics, and he introduced this subject at the University level in 1882. He was elected a member of the French Academy of Medicine at the age of forty-eight and a Deputy in the French Parliament after the First World War. He promoted the proposal for a premarital medical examination but was unsuccessful. He died in 1934.

Pinard disapproved of Lapouge who was a Darwinist: as a Lamarckian he believed in the role of education and environment inasmuch as the progress made in one generation was supposed to benefit the following generation. This is the reason why it is not sufficient merely to choose parents. One must also improve the standards of life and especially the conditions of procreation. As he wrote in 1903:

"The procreation act is the only instinct that has not been civilised. The highest and most valuable act a man can perform during his life which commands the conservation and progress of the race is still performed at the beginning of the twentieth century as it was in the stone age"[20].

In 1898 he gave his first lecture on "Conservation and Improvement of Race" at the Clinique Baudelocque. He defined his conception of eugenics without using the term and while quoting the term "puériculture" the word coined by Caron in 1865, he added:

"When practising puériculture (childcare) before procreation i.e. in doing prophylaxy, one will manage to lower the numbers of unfits, abnormal people, idiots and degenerates. The future of race is by and large dependent upon puericulture before procreation" and he stressed the specificity of "puériculture":

"A new science, French in its origins, which strives to discover and to apply the useful knowledge for developing, conserving and improving the human race. It is the result of three components:

1) puériculture before procreation i.e. eugénnétique

2) puériculture from procreation to birth

3) puériculture after birth"

Pinard was responsible for a discovery of the utmost importance at the Clinique Baudelocque: he noted that babies whose mothers used to relax before giving birth were in better health than the others.

In 1912 he coined the term "eugennétique" to define the science of the good conditions of procreation.[21] He was more confident in the improvement of the standards of life than in the selection of mates and especially in coercive measures to eliminate the unfit from procreation because such decisions led to a lower birth rate.

The third trend of French eugenics is to be found in neo-Malthusianism. The neo-Malthusians were looking for the "improvement" and the "progress" of humanity. This goal implied a control of procreation i.e. what they called "conscious procreation". In other words they wanted to substitute a "rational" or "artificial" selection for the natural one which had been perverted insofar as natural selection was halted by the process of civilisation.

The neo-Malthusian movement appeared at the end of the nineteenth century when Paul Robin founded the League for Human Regeneration in 1896.[22] Robin was born in 1837 in a family of the middle bourgeoisie of South France. He was a student at the Ecole Normale Supérieure and became a Professor of Natural Sciences in different high schools. All in all he was a revolutionary militant, a member of the Internationale and he took part in the Committee of the Internationale as an aide to Karl Marx. After the defeat of the French Commune in 1871 he left France for England where he earned his living as a teacher.

Having been granted amnesty he returned to France in 1880 and then dedicated himself to education as the Director of the orphanage of Cempuis in the department of Oise. There he favoured what he called "integral education" and pioneered coeducation, outdoor teaching and visits to workshops and factories. He opposed patriotism, religion and the dogmatic teaching of morals. In 1894 he was dismissed after violent attacks of right wing oriented newspapers which accused him of moral depravation and the corruption of youth.

He was now free for his last battle: the spread of neo-Malthusianism in France.

He was responsible for the introduction in France of the ideas of George Drysdale whose book *The Elements of Social Science* he considered as "the Bible of Humanity"[23] and he devoted himself to propaganda among the working class. In 1902 he summed up his ideas in a kind of motto: good birth, good education, good social organisation. He committed suicide in 1912.

The neo-Malthusians were not only efficient propagandists using all kinds of techniques to convince the workers but they also retailed condoms and contraceptives with directions for use. The Alliance Nationale contre la Dépopulation denounced them and launched against them violent attacks for "pornography" and "obscenity". Some of them were condemned and put in jail.[24]

The neo-Malthusians had adopted Malthus and especially his "law" stating that the geometrical progress of population and the arithmetical increase of food supply would result in a catastrophic situation if no limits were fixed.

From their point of view the concern for quality must always prevail. It was connected with the question of degeneracy and regeneration of the race. Birth control implied a limitation of the birth rate among the "degenerate people", the "inferior" who should avoid reproducing. Paul Robin favoured sterilisation of the degenerate:

"For the worst degenerate who cannot be cured there is no other remedy than artificial sterilisation"[25]

The limitation of natality was also necessary to solve the "social problem" i.e. the question of the relation between the proletariat and the capitalist class. First the fewer the children the better the education that can be provided and secondly the fewer the workers the better the wages because the capitalists will be obliged to pay more.

In spite of their differences these trends converged at more than one point. What is common to them lies in the historical context of the development of the French eugenic movement. In the second half of the nineteenth century the French were obsessed by what they called "dépopulation" which never took place.[26] What was true was the lowering birth rate and the fact that the mortality rate happened to be higher than the birth rate. This fear reached a peak after the war of 1870 with Prussia and the French defeat. It was paramount in the years preceding the First World War.

A second common feature of French eugenics is to be found in neo-Lamarckism and especially in the inheritance of acquired characters. It is to be noted that the neo-Lamarckians held dominant positions at the University level during the first half of the twentieth century.[27] This is why genetics developed so late in France - after the second World War - and why the faith in education, combined with meritocratic ideas, flourished in France among the neo-Malthusians as within the pro-natality movement.

The ubiquity of the "dépopulation" issue was an obstacle to the emergence of a "negative" eugenics which would have implied sterilisation, castration and elimination of the degenerate and could have influenced the birth rate.[28] Although some French eugenicists proposed sterilisation and castration for the degenerate (Dr Binet-Sanglé in his book *The Human Race*, 1918[29] and Dr Charles Richet in *The Human Selection*, 1913, published in 1919), the majority of the French eugenicists were opposed to such measures.

Another major issue lies in the racial problem. Despite their differences the French eugenicists were united in their belief that there is no "pure race", no French race (in spite of the title of Dr Martial's book[30]) inasmuch as France has always been a country of migrations and of mixed populations. Concerning immigration, the majority of them were attached to the idea of the "ethnical complex" a notion defined in the book by Eugène Pittard, *The Races and History* (1924).[31]

Immigration was necessary to guard against "dépopulation" but it presupposed the selection and the choice of immigrants. Top priority should be given to Europeans - i.e. to those people who are the closest to the French from a cultural point of view. Secondly immigration must provide France with industrial and land workers and future citizens. Immigration has been closely linked with assimilation and integration of foreigners.

Under such circumstances and despite their differences, most of the French eugenicists believed that as far as population was concerned, instead of being opposed, quality and quantity were interwoven because quality would emerge from quantity. These views are those of eugenicists and demographers such as Charles Richet, Lucien March, Alexis Carrel, Edouard Toulouse, Adolphe Pinard, Adolphe Landry, Alfred Sauvy among others.

One can understand why the programme of eugenicists merges to some extent with that of hygiene, which stemmed from Pasteur and claimed to be an applied social science. The hygienists denounced what they called "the social evils" i.e. alcoholism, syphilis and tuberculosis. As social reformers they were promoting a public health policy to improve the race and the human capital. They were also convinced that education was the best tool to achieve the social reforms that they wanted to implement. That faith in education was common to the eugenicist programme.

As the well-known eugenicist Sicard de Plauzoles wrote in 1945:

> "Eugenics needs before all the reform of habits and a moral reform, the respect of hygienic and moral rules, the implementation of a policy oriented against the social evils: slums, alcoholism, prostitution, syphilis"

and he added:

> "Referring to Frederic Houssay one can say that French eugenics will not be established by constraint but by

individual conscience. That eugenics will be practised depends upon those who procreate, of their education and voluntary action.[32] Consequently eugenics is based on:

- moral education
- free selection through the marital examination
- sexual education of those who procreate"[33]

Education is the central belief of French eugenics: we find in the neo-Malthusian books the notion of "eugenic education", that of "puériculture" by Pinard; "education" and "meritocracy" in Carrel's work; "moral education" by Sicard de Plauzoles.

Secondly the eugenicists like the hygienists were fighting against the "social evils" during the twentieth century. It must be stressed at this stage that the same people could be members of hygienist and eugenist societies. For example, as already mentioned, Sicard de Plauzoles was a member of The League for Human Regeneration, of The French Society of Eugenics founded in 1913, of The French League for Human Rights and Director of the most important centre of study of venereal diseases in France The Institute Alfred Fournier.

It is now difficult to imagine after Nazism that a left-oriented, radical and revolutionary eugenics existed before World War I but this was so and not only in France. However the historical context of the development of eugenics in France was responsible for its emphasis on positive eugenics.

Finally it is clear that the notion of a "French eugenics" mentioned by Sicard de Plauzoles in 1945 refers to these features: the importance of positive eugenics, the battle against the social evils and the reference to education as the best way to eugenics.

When we turn to the practical outcome one must recognise that eugenics never played a great direct role in France. First the movement was never strong enough. The institutionalisation has always remained limited. With one hundred members at its start in 1913 - a majority of these were doctors - the French Eugenics Society quickly declined after World War I[34] and lost its autonomy in the thirties to merge with the Ecole d'Anthropologie. Secondly the Catholic Church was opposed to eugenics inasmuch as the control of reproduction cannot depend upon human will.[35]

Nevertheless one can point to some achievements of eugenics in France.

In 1935 Alexis Carrel published his best seller *Man the Unknown* which contributed to the spread of his eugenic ideas. The success of the book was immediate, undeniable and lasting as is shown by the printing of more than a million copies in French from 1935 to 1980, translations into more than twenty languages and sales continuing to the present day. His eugenics is "positive", natalist and "voluntary":

"Eugenics is indispensable for the perpetuation of the strong. A great race must propagate its best elements ...

Eugenics must exercise a great influence upon the destiny of the civilised races. Of course the reproduction of human beings cannot be regulated as in animals. The propagation of the insane and the feeble-minded, nevertheless, must be prevented. A medical examination should perhaps be imposed on people about to marry, as for admission into the army or the navy, or for employees in hotels, hospitals and department stores ..."

It seems that eugenics, to be useful, should be voluntary ...

"The free practice of eugenics could lead not only to the development of stronger individuals, but also of strains endowed with more endurance, intelligence, and courage. These strains could constitute an aristocracy, from which great men would probably appear. Modern society must promote, by all possible means, the formation of better human stock ..."[36]

But one should not forget the gap between theory and practice, especially between the ideas expressed in the book and the practical achievements of the Foundation.

Even though it was called a foundation, the French Foundation for the Study of Human Problems, whose Regent he was from 1941 to 1944 in occupied France, was not a charity. The Foundation, known more popularly as the Carrel Foundation, was created as a financially autonomous public establishment with full legal status and had a double mission:

- to study "all possible means of safeguarding, improving and developing the French population" and

- "to synthesise efforts undertaken by its own members or by others and to develop the science of man"

It is not surprising that the passing of time and painful feelings associated with an organisation created by the Vichy Government have contributed to its being forgotten and discredited insasmuch as in recent years the Front National referred to Carrel as a forerunner of ecology and as a "spiritual guru". As a consequence a controversy over his very name has been taking place in France leading to its being removed from street names in many cities and also from the Faculty of Medicine in Lyon, his birthplace, at the end of 1995.

However the time has come to reassess its role especially from the point of view of the history of social sciences in France. The variety of its undertakings, their importance, the work achieved and the diversity of its effects cannot be left unrecognised forever.

In January 1944 there were more than two hundred and fifty people working for the Foundation. The initial budget of forty million francs was almost equal to that of the National Center for Scientific Research, and its offices and laboratories all over Paris and the surrounding areas, especially at Meudon Bellevue, highlights the importance of the resources Carrel had at his disposal.

While striving to encourage a spirit of synthesis and "collective thought", the Foundation brought to France empirical and experimental research methods used in the United States, contributing to what could be described as a methodological transplant.

Considering its short official existence and the exceptional circumstances of the period, the scientific work of the Foundation is impressive: the demographic analyses undertaken by Robert Gessain, Paul Vincent and Jean Bourgeois, the pioneer work of Jean Sutter in nutrition, Jean Merlet's work on group settings, the polls undertaken by Jean Stoetzel's team and the publications of the François Perroux Department of Biosociology.

Amongst the achievements of the Foundation one should mention the National Institute for Demographic Studies and with it the establishment of one of the most active and productive research groups in the social sciences in France. There are other initiatives and activities which are part of the Foundation's legacy: the creation by Dr. André Gros, the former Vice-Regent of the Foundation of the group of "Advisers of Synthesis" in 1947; the joint creation by Dr Gaston Berger of the group and the publication "Prospective" ten years later; the wide and almost unique multidisciplinary study undertaken in 1960 by the "Délégation Générale à la Recherche Scientifique et Technique" thanks to the initiative of Robert Gessain and Jean Sutter on a commune of Britany Plozévet and the Monaco Forums on social sciences. We should also mention the contribution of former Foundation members to the establishment at the national and international level of occupational medicine and the encouragement given to ergonomic studies.

In the field of demography let us first mention the article published by Jean Bourgeois in the November 1945 issue of the *Cahiers de la Fondation française pour l'étude des problèmes humains* establishing the renewal of fertility after 1941-1942; this was so surprising that nobody believed it.[37] Secondly the statistical inquiry called the "100,000 children inquiry" because this was the size of the sample defined by Jean Stoetzel as necessary and which is the matrix of all subsequent research in the sociology of education in France.[38]

Concerning eugenics the only practical measures taken in France were the premarital medical examination, the health care manual created by a law passed on 16 December 1942[39] and the creation of occupational health by a law passed on 23 July 1942. We must bear in mind that these two laws were not repealed after the Liberation of France. Even if there is no evidence that the Carrel Foundation played a direct role in the legal process, it helped and supported it.

The limelight goes to occupational medicine for which Alexis Carrel provided inspiration and developed the philosophy - essentially the cornerstones of his work - solidarity and the resolution of social conflicts. The law of 23 July 1942 which made it obligatory to have medical and social services in a factory was the result of a long historical process to which doctors of the Foundation made a significant contribution, both during the war and at the Liberation.

A pure institutional analysis could lead to a misunderstanding of the real influence of eugenics on the French society. As in other European countries eugenics influenced French policy on public health. Firstly, the creation of motherhood and childcare centres after World War II derives from the ideas of Pinard and from his concept of puericulture. Secondly, there is a direct relation between neo-Malthusianism and the movement for family planning that emerged after 1956. Thirdly, what is "therapeutic abortion" as defined by the law of 1975 if not a statement and a decision inspired by negative eugenics?

Eugenics also influenced the definition and scope of some social sciences: demography[40], psychology[41], physical anthropology whereas it essentially influenced statistics and sociology in Great Britain.[42] It must be emphasised that the history of relations between eugenics and social sciences in France has not yet been written. It was part of the "science of man" of Carrel and a fundamental of the biocratic idea of Toulouse[43] and Carrel which held that eugenics should contribute to the reconstruction of Western civilisation as a tool to select the good stocks of population.

In conclusion I would like to deal briefly with the relationship between the present bioethical debates in France and the history of eugenics.

Whereas in the last decades of the nineteenth century eugenics developed its scientific and progressive project of improving the human race without a real knowledge of human heredity, recent progress of biology and genetics have completely transformed the issues being discussed. We are now on the eve of a new era with the mastery of techniques of procreation which is giving birth to new fears among the population (a typical word used when talking about eugenics is the word risk or "dérive" in French)

Secondly, contemporary eugenics is no longer a pure intellectual issue: it is a popular one inasmuch as the new techniques of procreation are the answers to demands from infertile people.

Thirdly, actors concerned or involved with the debates are not used to meeting and to discussing together. Consequently prejudices and misunderstanding are the rule. Physicians and biologists ignore the history of eugenics and historians are not trained to understand the scientific data linked to the debates. Public opinion is not informed and the great majority is not able to make up its mind. There is a lack of democratic debate in France on bioethical issues. The French National Committee on Ethics – (the first in Europe to be officially created in 1983) is responsible for a two-day meeting every year with the public but beyond this event there is nothing which could be compared with what is done in other countries, especially in a Northern State like Denmark for debate-generating activities.

Nowadays eugenics is either ignored or condemned in France whereas it is practised every day. This is the great paradox and contradiction of the situation in present-day France. First it is practised through the selection of sperm donors. Secondly we should not forget the so-called "therapeutic abortion" defined by the law of 1975 on contraception. What is dominant is opacity. The statistical data are too limited and contain too much bias to allow any assessment of the efficiency of the techniques. Nobody knows what a good gene is and consequently it is impossible to define the criteria of a voluntary orthogenics. We can consequently distinguish two different attitudes: one favours abstention, the other defends the idea of a selection of donors (the idea of a bank of sperm of Nobel Prize winners was first presented by J.H. Muller; Julian Huxley favoured "eugenic insemination by deliberately preferred donors")

Who would and could define the best human type?

Among the arguments used in the debate we must remember the diversity of the human race that should be preserved and could be threatened by eugenics. The practical argument is also to be mentioned: beyond the fact that it is impossible to sterilise all the individuals in order to eradicate hereditary diseases, there is also a contradiction between the democratic creed and eugenics as Ernst Mayr noted:

> "To assert that human beings are genetically different, even if it was proved by science, cannot be accepted by the majority of public opinion in the West. There is an ideological gap between egalitarianism and eugenics".[44]

It must be added that the limits of knowledge combined with those of the future evolution of techniques mean only that man will never be able to achieve a total mastery of procreation.

Last but not least there is a philosophical argument: the practice of eugenics is at odds with the Kantian principle: never consider humanity as a means but as a goal.

If the belief in a total mastery of procreation and improvement of the human race is a pure fantasy, it does not mean that it is impossible to control the genetic stock. We are now confronted with a double bind: either nothing can be done or everything is possible.

Eugenics in Scandinavia

We now come to the case of Scandinavia. Instead of giving a detailed analysis of the eugenics movements in the four countries: Denmark, Sweden, Norway and Finland, I will try to highlight the main issues connected with the development of eugenics inasmuch as they illustrate the complexity of relations between science, ideology and politics.

What is at stake in this part of Europe is the question of welfare state eugenics i.e. the relation between the creation of the Welfare State in these countries and eugenics. My analysis refers to the studies gathered by Gunnar Broberg and Nils Roll-Hansen in a book published in 1996 and entitled *Eugenics and the Welfare State*.[45]

Despite many studies some points still remain arcane: for example in Denmark, the first country of Scandinavia to pass a sterilisation law in 1929, eugenics has been practised without institutionalisation of eugenics: there was neither a eugenic society nor a teaching of eugenics at the University level and the term eugenics never appeared in the legislative process. The report of the Commission which prepared the first law -the law of 1929 - was entitled "Social measures toward degeneratively predisposed individuals". In other words practice does not depend on the existence of a eugenic movement and vice versa the institutionalisation of eugenics does not always lead to implementation. At the European level comparative analysis could be fruitful to help to understand the genesis of eugenic policy.

In the northern States of Europe two waves of eugenics took place: the first one just before WW1, the second one in the 1930s and 1940s. In these countries eugenics was a significant issue of social policy and there was extensive public interest in the subject even though eugenics organisations were weak. Sweden was the only country with a national eugenics society - the Swedish Society for Racial Hygiene was founded in 1909[46]; in the others various organisations with social causes like the Association of Public Health in Swedish speaking Finland took on some of the same tasks. There were groups of active people doing propaganda such as Mjöen's Consultative Eugenics Committee of Norway.

Even if the organised movement was most visible before the First World War, the establishment of Herman Lundborg's Institute for Racial Biology in Uppsala in 1922 is an essential date for institutionalisation of eugenics. It was the first State Institute of that kind in the world established with a staff of seven persons and an initial budget of 60,000 crowns. Its first major work was the publication in English of *The Racial Characters of the Swedish Nation* (1926). But it was during the second wave that eugenics was the most effective with the sterilisation laws of the 1930s.

The first wave is grounded in physical anthropology, racism with a reference to the Nordic race; the second developing in the 1920s and the 1930s included antiracist positions and required a better knowledge of genetics.

The close link between eugenics and the movement for social reforms is well established by Gunnar Broberg and Nils Roll-Hansen. Eugenic sterilisation was an integral part of the social welfare state that emerged in the 1930s and the 1940s insasmuch as it would reduce the cost of institutional care, special schools and poor relief.

Let us take two examples: one in Denmark and the other one in Sweden.

In Denmark the central political figure is K K Steincke, a long time minister of Justice, later of Health and Welfare, and who claimed in the name of the Social Democrats the need for eugenics. The political trade-off was as follows: in a civilised society one must take care of the handicapped, of the mentally retarded, of the feeble minded. One must provide them with the best comfort in order to help them in supporting their handicaps but on one condition: they should not reproduce. Consequently he favoured the sterilisation laws of the thirties.

Concerning Sweden. The Myrdals - Alva and Gunnar - who inspired the Swedish population policy in the thirties defended eugenics for two major reasons: first it is part of the necessary "process of social adjustment" to industrial society and secondly it could help in financing the spread of welfare in society by lowering the cost of social relief:

"In our day of highly accelerated social reforms the need for sterilisation on social grounds gains new momentum. Generous social reforms may facilitate homemaking and childbearing more than before among the groups of less desirable as well as more desirable parents. This may not be regretted in itself as the personal happiness of these individuals and the profitable rearing of those of their children already born are not to be neglected. But the fact that community aid is accompanied by increased fertility in some groups that are hereditarily defective or in other respects deficient and also the fact that infant mortality among the deficient is decreasing demands some corresponding corrective"[47]

The Scandinavian experiences also show clearly how eugenics is not apart from scientific research and how science is part of the social process of the construction of the Welfare State. Some of the most prominent geneticists, among them Wilhelm Johannsen[48], Tage Kemp[49] in Denmark, Herman Nilsson-Ehle[50] in Sweden, Harry Federley[51] in Finland while doubting and questioning eugenics contributed to the passing of the laws of sterilisation either as members of the commissions which prepared the bills or as experts of the committees which took the decisions.

Let us turn now to the sterilisation laws and their implementation.

All the Nordic sterilisation laws of the 1930s assumed that permission for sterilisation had to be given by government authorities. It was illegal to perform sterilisation without permission. Consequently, when looking at the sterilisations, one must distinguish two periods: before the Second World War and after.

During the first period the number of sterilisations in the Nordic countries is relatively low: 108 in Denmark from 1929 to 1935 and 1380 during the next five years; 3000 in Sweden from 1935 to 1941 (first sterilisation Act); around 1000 in Norway from 1934 to 1945; 1908 from 1935 to 1955 in Finland.

Things changed at the end of the war inasmuch as sterilisations were more and more combined with abortions.

Today under the new sterilisation laws introduced in the 1970s the principle of individual freedom now prevails. This has resulted in a large increase in the number of registered sterilisations in Sweden. The number had reached a level of about 2400 around 1950 and then declined to 1500 in 1974. When the new sterilisation law was introduced in 1975 it rose to nearly 10,000 in 1980.

Prevention of sexual crimes was a motive for sterilisation. Women's organisations were particularly active in sending petitions to support the sterilisation laws, often referring to the need for preventing sexual crime. The first Danish law passed on 1 June 1929 distinguished two different groups of individuals. Section 1 aims at sex offenders:

"Persons who through the abnormal degree or character of their sexual desire are liable to commit crimes"

Section 2 deals with the question of offspring:

"Operations on the genital organs may be permitted in psychically abnormal persons, who, although they do not present any such danger to the public security as dealt with in Section 1 still render it important to society and to themselves that they be rendered incapable of reproduction"[52]

It is a sign of the importance of the connection to crime prevention that sterilisation applications in Sweden were handled by the Board of Health's Forensic Psychiatric Committee until 1947 ... Still the numbers of sterilisations carried out on sexual criminals was small in all Nordic countries. Relatively few castrations were carried out, most of them in Finland where eugenic sterilisation in general came later and lasted longer then in other countries (totalling 90). In all four Nordic countries the great majority of eugenic sterilisations were on the mentally retarded.

The predominance of women among the sterilised - in the Swedish case more than 90 percent - must also be underlined ... the implementation of the sterilisation laws was not gender neutral. Further research is necessary to explain how and why women became the primary object of the Scandinavian sterilisation programs.

In Denmark, Sweden and Norway the number of sterilisations of the mentally retarded and insane dropped from the middle of the 1940s through the 1950s. But there is no evidence that the decline of eugenics is the consequence of revulsion caused by the Nazi crimes.

Those who promoted eugenics in the 1930s still supported it after 1945. This was certainly the case with geneticists of a liberal or socialist bent, such as J Huxley, J B S Haldane in England, or J H Muller in the US, or H. Nachtsheim in Germany. In Scandinavia key medical or biological experts like Tage Kemp (Denmark), Nils von Hofsten (Sweden), C.A. Borgstôm (Finland) and Karl Evang (Norway) all supported sterilisation of the mentally retarded well into the 1950s.

Decrease in eugenic sterilisation did not result in a decrease in the total number of sterilisations.

First there was change from eugenic indications to medical indications.

One must remember that physicians had a professional interest in eugenics and played a central role in developing its policies and practices. It is the case in Denmark where 80% cent of the sterilisations (all of women) were performed on the initiative of doctors with the help of Mother's Aid Agencies. 70% of sterilisations were combined with abortion and 80% connected with a pregnancy.

Second there was a trend toward using sterilisation as a means of contraception. The steadily increasing number of sterilisations in Norway during the post-war period until the 1970s is due to this development. When a new Norwegian law was introduced in 1977 explicitly stating that for any person with full legal rights sterilisation was a matter between patient and doctor, that was only the normalisation of a practice already instituted. In Sweden a less liberal attitude seemed to have prevailed. The Swedish Population Commission in 1936 found the idea that every person should be free in all respects to determine the use for his or her own body to be "an extremely individualistic view".

From an international perspective the comparison of the Nordic countries with Germany is particularly interesting. In these countries also the question of the continuity is the major issue at stake.

The relations between the Danish eugenicists and the German eugenicists were ambiguous: while preferring the Danish eugenics policy to the Nazi one, Tage Kemp helped after World War II some of the prominent figures of Nazi eugenics such as Otmar von Verschuere and Fritz Lenz to join the scientific community.

Finally the decline of eugenics in Scandinavia appears to have other causes beyond revulsion against the events in Germany.

The welfare state emerged in countries that had all had Lutheran state churches, a relatively homogeneous culture and a relatively egalitarian social structure. Strong labour parties co-operated with strong labour organisations and were winning government power that was to last more or less continuously for the next half century.

The progress of the knowledge of genetics, in particular human genetics from the early years of the twentieth century was a prerequisite for the political debates and decisions. But there was also an underlying view of the relation between science and politics which linked eugenics to the development of the welfare state that was so typical of the Nordic countries in the middle decades of the twentieth century. In continuation of the enlightenment view of science, social and economic planning based on science was seen as the motor of social progress. These views are now being revised inasmuch as the welfare state itself is facing a fundamental crisis.

Table 1: Reported Sterilisations in Sweden, 1935-1975.

Year	Eugenic indication	Social indication	Medical indication	Total	Percent women
1935	-	-	-	250	94
1936	-	-	-	293	93
1937	-	-	-	410	91

Table 1: Reported Sterilisations in Sweden, 1935-1975.

Year	Eugenic indication	Social indication	Medical indication	Total	Percent women
1938	-	-	-	440	93
1939	-	-	-	523	94
1940	-	-	-	581	83
1941	-	-	-	746	69
1942	959	67	135	1,161	63
1943	1,094	52	181	1,327	65
1944	1,437	21	233	1,691	65
1945	1,318	78	351	1,747	73
1946	-	-	-	1,847	-
1947	1,210	65	845	2,121	86
1948	1,188	53	1,023	2,264	87
1949	1,078	44	1,229	2,351	91
1950	858	17	1,473	2,348	94
1951	629	48	1,657	2,334	95
1952	405	73	1,635	2,113	95
1953	330	75	1,434	1,839	96
1954	204	72	1,571	1,847	96
1955	159	76	1,602	1,837	97
1956	172	76	1,520	1,768	97
1957	149	90	1,546	1,785	97
1958	-	-	-	1,786	96
1959	-	-	-	1,849	95
1960	75	120	1,455	1,650	96
1961	62	118	1,619	1,799	96
1962	33	94	1,558	1,685	98
1963	48	96	1,605	1,749	97
1964	34	70	1,655	1,759	98
1965	11	22	1,475	1,508	99
1966	9	26	1,500	1,535	99
1967	1	42	1,465	1,508	99
1968	13	20	1,545	1,578	99
1969	19	58	1,496	1,573	99
1970	20	46	1,797	1,863	99
1971	13	63	1,826	1,902	99
1972	12	45	1,559	1,616	99
1973	17	19	1,358	1,364	99
1974	21	6	1,487	1,514	99
1975	14	3	1,011	1,028	99
1935-1975				62,888	93

Source: *Sveriges Offentliga Statistik: Allmän hälso- och sjukvård* (Stockholm: Statistika centralbyrån 1935-1976) [Annual reports on health published by the Swedish Central Bureau of Statistics.]

Figure 1: Reported Sterilisations in Sweden, 1942-74, and Indications

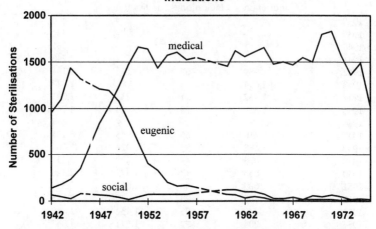

Source: *Sveriges Offentliga Statistik: Allmän Hälso- och sjukvård* (Stockholm: Statistika centralbyrån 1935-1976) [Official Statistics of Sweden: Health]. Note: no infomation available for 1946 and 1958-59.

Table 2: Number of Sterilisations with Permission of the Finnish Board of Health, 1935-1955

Year	Total	percent of those women
1935-36	54	85.2
1936-37	102	76.5
1937-38	121	81.0
1938-39	112	72.3
1939-40	32	65.6
1940-41	37	78.4
1941-42	27	74.1
1942-43	24	91.7
1943-44	42	88.7
1944-45	37	83.8
1945-46	67	89.6
1946-47	84	91.7
1947-48	73	89.0
1948-49	82	85.4
1949-50	102	86.3
1950-51	189	92.1
1951-52	136	94.1
1952-53	162	87.0
1953-54	201	86.1
1954-55	224	86.2

Source: C.A. Borgström, *Tillämpningen av lagen om sterilisering i Finland 13.6.1935-*

Table 2: Number of Sterilisations with Permission of the Finnish Board of Health, 1935-1955

Year	Total	percent of those women

30.6.1955 kasteringarna obcaktade, Bidrag till kännedom av Finlands Natur och Folk 103 (Helsingfors: Findska Vetenskapssocieteten, 1958), 50.

Table 3: Sterilisations in Finland, 1951-1970.

Year	Total	Medical Reasons*	Eugenic Reasons**	% of total	Social Reasons**	% of total
1951	780	569***				
1952	1009	777***				
1953	1061	813***				
1954	1068	733***				
1955	1236	1014***				
1956	1582	1582	362	22	23	1
1957	1727	1727	372	22	21	1
1958	2206	1767	413	19	26	1
1959	2596	1921	436	17	239	9
1960	3197	2362	514	16	321	10
1961	3193	2353	463	15	377	12
1962	3388	2612	411	12	365	11
1963	3511	2729	380	11	402	11
1964	3297	2676	216	7	405	12
1965	3206	2713	237	7	258	8
1966	3543	3012	271	8	260	7
1967	4022	3521	269	7	232	8
1968	4294	3817	218	5	259	6
1969	5437	4983	298	5	156	3
1970****	5727	2385	141	5	101	3
total	56,080	44,066	5001		3445	

Source: Public Health and Medical Care, The Official Statistics of Finland XI, 1950-1970

* Either based on a consensus decision of two doctors or on the evidence of somatic disease or defect as agreed by a statement of the National Board of Health.
** Statement of the National Board of Health.
*** With authorisation of two doctors.
**** 1.1-31.5 Based on sterilisationlaw of 1950; 1.6-31.12 Based on new sterilisation law of 1970.

Table 4: Sterilisations in Norway, 1934-1976, Granted Applications.

Period	Number of Sterilisations	Percent Women	Annual Average

1 June 1934-31 December 1942	653	83	76
1 January 1943-8 May 1945 (Nazi law)	487	84	207
9 May 1945-30 June 1954	2,569	91	283
1 July 1954-1965	8,005	93	696
1996-1976	29,177	62	2,652
1934-1976	40,891	ca 75	951

Source: K. Evang, *Sterilisering etter lov av 1. juni 1934 om adgang til sterilisering m.v.* (Sarpsborg: F. Varding, 1955), 13; The Norwegian Parliament, Government Bill 1976/1977 no 46, "Om lov om sterilisering m.v.," 16.

Note: No figures for 1 July-31 December 1959. Sterilisations on medical indications are not included. Higher figures during the Nazi law, some 280 annually, are estimated by Gogstad (A. Gogstad, *Helse og Hakekors. Helsetjeneste og helse under okkupasjonsstyret i Norge*, 1940-45 [Bergen: Alma Mater Forlag, 1991], 209.

Notes and References:

[1] Alain Drouard, *Une inconnue des sciences sociales La Fondation Alexis Carrel (1941-1945)*, Paris, Ined, Editions de l a Maison des sciences de l'homme, 1992

-*Alexis Carrel (1873-1944) De la mémoire à l'histoire*, Paris, L'Harmattan, 1995

Carrel' best seller has been published under the title: *L'homme, cet inconnu* (Paris, Plon, 1935) and *Man, the Unknown* (New York and London, Harper & Brothers), 1935.

[2] *Inquiries into Human Faculty and its Development*, The Eugenics Society, p 17.

Galton gave other definitions of eugenics such as:

"the study of agencies under social control that may improve or impair the racial qualities of future generations" or in 1907:

"Eugenics is the science which deals with all influences that improve the inborn qualities of a race, also with those that develop them to utmost advantage"

[3] P.A. Taguieff, L'eugénisme, objet de phobie idéologique, *Esprit*, 11, novembre 1989.

[4] Adams, Mark B., ed. *The Wellborn Science: Eugenics in Germany, France, Brazil and Russia*, New York, Oxford University Press, 1990, 242p.

[5] The concept of "scientific ideology" has been defined by Georges Canguilhem in a conference entitled Qu'est-ce qu'une idéologie scientifique? published in *Idéologie et rationalité dans l'histoire des sciences de la vie*, Paris, J. Vrin 1977, on pages 33-45 .

[6]Muller, Hermann J., *Out of the Night: a Biologist's View of the Future*, New York, Garland Pub., 1984, c 1935, 127p.

[7]G. Vacher de Lapouge, "L'anthropologie et la science politique", *Revue d'anthropologie*, 15 mars 1887.

[8]Anne Carol, *Les médecins français et l'eugénisme 1800-1942 De la mégalanthropogénésie à l'examen prénuptial*, Thèse pour le doctorat d'histoire, Université de Paris I, Panthéon-Sorbonne, 1993. This thesis which has been published under the title *Histoire de l'eugénisme en France Les médecins et la procréation XIXe -XX e siècle*, Le Seuil, 1995 shows the importance of the pre-Galtonian medical eugenics in France .

[9]Charles Richet, *La sélection humaine*, Paris, Félix Alcan, 1919.

[10]Rostand, Jean, *L'homme* Introduction à l'étude de la biologie humaine, NRF, Gallimard, L'avenir de la science, 1926.

- *Instruire sur l'homme*, La Diane française, 1953.

[11]Among the recent studies on French eugenics let us mention: Schneider, William H., *Quality and Quantity: The Quest for Biological Regeneration in Twentieth-Century France*, New York, Cambridge University Press, 1990, 392p

[12]Prosper Lucas, *Traité philosophique et physiologique de l'hérédité naturelle*, Paris, J.B. Baillère, 1847-1850,(2 vol.)

[13]A. Caron, Introduction à la puériculture et à l'hygiène de la première enfance, Paris, l'auteur, 1865.

[14]Henri de La Haye Jousselin, *Georges Vacher de Lapouge (1854-1936) Essai de bibliographie*, Paris, 1986. From the same author: *L'idée eugénique en France Essai de bibliographie*, Paris, 1989.

[15]Letter dated June 20,1927 in Archives of the Institute for Racial Biology of Uppsala.

[16]Among social darwinists, one can also mention Clémence Royer (1830-1902) who first translated Darwin into French and introduced his work in France and the Nobel Prize winner in medicine Charles Richet (1850-1935). More precisely Charles Richet combined Lamarckism and Darwinism.

[17]Georges Vacher de Lapouge, *Les sélections sociales* Cours libre de science politique professé à l'Université de Montpellier (1888-1889), Paris, A. Fontemoing, 1896, p262.

His two other major books are: *L'Aryen, son rôle social* Cours libre de science politique à l'Université de Montpellier, 1888-1889, Paris,

Fontemoing, 1899, and *Race et milieu social* Essai d'anthropologie, Paris, Marcel Rivière, 1909.

[18] Vacher de Lapouge, "La race chez les populations mélangées" in *Eugenics in Race and State*, vol.2, Williams & Wilkins Company, 1923, p.6.

[19] Lefaucheur Nadine, La puériculture d'Adolphe Pinard in M. Manciaux et G. Rambault eds. *Enfance menacée*, Editions de l'Inserm, 1991.

[20] Pinard's answer to an inquiry published in *La Chronique médicale*.

[21] "..Eugénnétique aims at studying and diffusing the best conditions of reproduction" in "De l'eugénnétique", *Annales de gynécologie et d'obstrétique*, décembre 1912.

[22] Demeulenaere-Douyère, C., Paul Robin (1837-1912) Un militant de la liberté et du bonheur, Published, 1994.

[23] George Drysdale, *Eléments de science sociale par un docteur en médecine*, 6ème édtion française d'après la 32ème anglaise, Librairie malthusienne, 1905.

[24] During the years 1907-1908-1909 Paul Robin, Eugène Humbert, Louis Grandidier, Dr Liptay, Dr Elosu, Lerouge, Liard Courtois, Hureau, Cauvin, Fabry were condemned or sent to jail.

[25] Paul Robin, *Le néo malthusianisme*, Librairie de Régénération, 1905.

[26] The issues connected to "dépopulation" were raised by many authors and writers during that period. We may refer to Prévot-Paradol, Elisée Reclus, Vacher de Lapouge, Arsène Dumont. Many books were published on this issue: Arsène Dumont, *Dépopulation et civilisation,* Paris, Le Crosnier, 1890; Georges Rossignol, Un pays de célibataires et de fils uniques, 1896 (first edition); Librairie Ch. Delagrave, 1913; Emile Levasseur, *La population française*, 1889-1892, vol.III,; Dr Jacques Bertillon, *La dépopulation de la France*, Alcan, 1911.

[27] Almost all the professors of zoology, biology and natural sciences at the Sorbonne or at the Museum - let us mention the names of Maurice Caullery, Edmond Perrier, Yves Delage - were neo Lamarckians; others tried to combine Lamarckism and Darwinism as Charles Richet did or neo Lamarckism and Mendelism as Alexis Carrel. Lucien Cuénot stands apart.

[28] These views are clearly expressed in Alfred Fabre-Luce's book: *Pour une politique sexuelle*, Grasset, 1929.

[29] Binet-Sanglé, Dr, *Le haras humain*, Paris, A. Michel

[30] Paul Broca who founded the physical anthropology in the middle of the nineteenth century was already expressing this view in a conference

entitled "Sur la prétendue dégénérescence de la population française" (1867, *Bulletin de l'Académie impériale de médecine*, tome XXXII, p547)

René Martial, Dr, *La race française*, Paris, Mercure de France, (first edition, 1934). René Martial specialised early in immigration problems and published numerous studies on this issue. Among them let us quote his most important book: *Traité de l'immigration et de la greffe interraciale*, Paris, Librairie Larose et Cuesmes-les-Mons (Belgique), 1931.

[31] Eugène Pittard, *Les races et l'histoire. Introduction ethnologique à l'histoire*, Paris, A. Michel, 1924. The heterogeneity and the mingling of races were at stake and defined as factors of political unity in Jacques de Boisgelin's book *Les peuples de la France Ethnographie nationale*, Paris, Didier et Cie, 1878.

[32] "L'eugénisme français" in *La prophylaxie antivénérienne*, 17ème année, n°10, octobre 1945. At the beginning of the XXth century the idea of a "national" eugenics is already to be found in the literature.

[33] Ibid.

[34] From one hundred members to around fifty in the mid-twenties.

[35] The conflict between eugenics and the Catholic Church is clearly analysed by Albert Valensin in *Hérédité et Races*, Les éditions du Cerf, 1931.

[36] *Man, the unknown*, 1935, Harpers and Brothers, on pages 299, 300, 302.

[37] Jean Bourgeois, "Evolution de la population françaize de 1939 à la fin de 1944", *Cahiers de la Fondation française pour l'étude des problèmes humains* n°4, novembre 1945.

[38] See among the studies published by the National Institute for Demographic Studies on this issues:

"Le niveau intellectuel des enfants d'âge scolaire. Une enquête nationale dans l'enseignement primaire présentée par le Professeur Georges Heuyer, le Professeur Henri Piéron, Madame Henri Piéron et Alfred Sauvy" (*Travaux et documents, Cahier n°13*, PUF, 1950.

"Le niveau intellectuel des enfants d'âge scolaire . La détermination des aptitudes. L'influence des facteurs constitutionnels, familiaux et sociaux. Analyses par René Gille, Louis Henry, Léon Tabah et Jean Sutter, Hélène Bergues, Alain Girard et Henri Bastide, préface de Henri Laugier" (*Travaux et documents, Cahier n°23*, PUF, 1953).

[39] Loi n° 941 du 16 décembre 1942 relative à la protection de la maternité et de la première enfance (Law of December 16, 1942 on the protection of motherhood and childhood). Beside the premarital medical exam the law

also created one "carnet de santé" (health care booklet) and required two medical exams from pregnant women. The law also defined the allowances given to pregnant women.

[40] At its beginnings the National Institute for Demographic Studies and its Director Alfred Sauvy defined demography as a social science which should be developed with a broad perspective including "qualitative" demography as well as quantitative demography. This concern with eugenics is also expressed in Jean Sutter's book *L'eugénique* (1951).

[41] One should here refer to the history of testing and psychotechnics in France from the beginning of the century with the Binet-Simon tests and to the efforts of Toulouse and Laugier to promote the scientific study of labour in France between the two world wars. See in W. Schneider's "The Scientific Study of Labour in Interwar France", *French Historical Studies*, - 1991 Vol. 17, n° 2.

[42] Donald A. Mackenzie, *Statistics in Britain 1865-1930* The social construction of knowledge, Edinburgh University Press, 1981.

[43] Edouard Toulouse was born in 1865 in Marseilles where he qualified as a doctor and a psychiatrist. He was appointed head psychiatrist at the asylum of Villejuif in 1898. He is responsible for a major reform of the asylum which took place in 1922 when he created an "open" sector at Sainte-Anne in Paris which gave birth later to the Hôpital Henri Rousselle. Edouard Toulouse also obtained a decree that changed the name of asylum into "psychiatric hospital". From Henri Sellier, Minister of Public Health.

He coined the term "biocratie" in 1920 in the review *Le Progrès civique*. From 1920 to his death in 1947 he considered the "biocratie" as the only solution of all human and social problems. See a thesis of medicine: Sage, Michel., *La vie et l'oeuvre d'Edouard Toulouse (aliéniste, psychologue et sexologue marseillais)*, Marseilles, Faculté de médecine, Thèse de doctorat de médecine, 1979.

[44] Ernst Mayr, *Histoire de la biologie Diversité, évolution et hérédité*, Fayard, 1989, on p.576. This book is the French translation of *The Growth of Biological Thought, Diversity, Evolution and Inheritance*, The Bellknap Press of Harvard University Press, 1982.

[45] Broberg, Gunnar and Nils Roll-Hansen eds., *Eugenics and the Welfare State Sterilization Policy in Denmark, Sweden, Norway and Finland*, Michigan University Press, 1996, 294p.

[46] At the beginning of the society zoologist Leche and anatomist Vilhelm Hultcranz were the key figures. Then Herman Lundborg joined and became an active member of the society.

[47] Alva Myrdal, *Nation and Family The Swedish Experiment in Democratic Family and Population Policy*, London, Kegan Paul, Trench, Trubner & co, 1945. As the author explains, the book is a new version for English speaking countries of the book written with Gunnar Myrdal and published under the title *Kris i befolkningsfragan* (Crisis in the population question).

[48] A major figure of Mendelian genetics, Wilhelm Johannsen (1857-1927) coined the terms gene, genotype and phenotype. In his book *Arvelighed i Historisk og Eksperimentel Belysning* (Heredity in historical and experimental light), published in 1917, he took position against the Galton eugenics, Pearson and the biometrical approach. He was also sceptical about positive eugenics.

[49] Tage Kemp (1896-1964) was the first Director of the Institute of Human Genetics of Copenhagen founded in 1938 with the support of the Rockefeller Foundation. After World War II he introduced the expression of "genetic hygiene" as a substitute to eugenics.

[50] A famous botanist Hermann Nilsson-Ehle was the first Professor of Genetics at the University of Lund.

[51] Harry Federley (1879-1951) is considered as the father of Finnish genetics. He was the first professor of genetics at the University of Helsinki in 1923 where he founded the Department of Genetics. Eugenics has always been part of his teaching.

[52] H.O. Wildenskov, Sterilization in Denmark A Eugenic as well a Therapeutic Clause, *The Eugenics Review*, vol.23, January 1932.

Eugenics In North America

Daniel J Kevles

As in Britain, eugenics in the United States and Canada had its roots in the social Darwinism of the late nineteenth century, with all its metaphors of fitness, competition, and inequality. A key proto-eugenic theme was that social measures interfered with natural selection and thus fostered the multiplication of the unfit, a trend that was said to lead to social degeneration. Proto-eugenicists on the western side of the Atlantic knew about and admired Galton. They took him as their patron saint, embracing his ideal of improving the human race by, as he put it, getting rid of the 'undesirables,' multiplying the 'desirables,' and encouraging human beings to take charge of their own evolution.[1]

In North America, Galton's eugenic ideas took broadly popular hold after the turn of the twentieth century. Adherents of eugenics were united by an absorption with the role of biological heredity in shaping human beings. Most eugenicists in the United states and Canada believed that human beings were determined almost entirely by their 'germ plasm,' their inheritable essence, which was passed on from one generation to the next and which overwhelmed environmental influences in shaping human development. Their belief was reinforced by the rediscovery, in 1900, of Mendel's theory that the biological makeup of organisms was determined by certain 'factors,' which were later identified with genes. Human beings, who reproduce slowly, independently, and privately, are disadvantageous subjects for genetic research. Nevertheless, since no creature fascinates us as much as ourselves, efforts were mounted and institutions established in the early twentieth century to explore human inheritance, especially eugenically relevant traits.

In North America, the most important such institution was the Eugenics Records Office, which was affiliated with, and eventually became part of, the biological research facilities that the Carnegie Institution of Washington sponsored at Cold Spring Harbor, on Long Island, New York, under the directorship of the biologist Charles B Davenport. Eugenic research included the study of the hereditary transmission of medical disorders - for example, diabetes and epilepsy - not only for their intrinsic interest but also because of their social costs. A still more substantial part of the program consisted of the analysis of traits alleged to make for social burdens--traits involving qualities of temperament and behaviour that might lie at the bottom of, for example, alcoholism, prostitution, criminality, and poverty. A major object of scrutiny was mental deficiency - then commonly termed 'feeblemindedness' - which was often identified by intelligence tests and was widely interpreted to be at the root of many varieties of socially deleterious behaviour. Typically for eugenic scientists, Davenport concluded that patterns of inheritance were evident in insanity, epilepsy, alcoholism, 'pauperism,' and criminality.

Such findings were widely disseminated in popular books, articles, and lectures, and they made their way into common culture. A chart displayed at the Kansas Free Fair in 1929, purporting to illustrate the "laws" of Mendelian inheritance in human beings, declared, "Unfit human traits such as feeblemindedness, epilepsy, criminality, insanity, alcoholism, pauperism, and many others run in families and are inherited in exactly the same way as colour in guinea pigs."[2]

Davenport helped introduce Mendelism into the influential studies of 'feeblemindedness' that were conducted by Henry H Goddard, the psychologist who brought intelligence testing to the United States. Goddard speculated that the feebleminded were a form of undeveloped humanity: "a vigorous animal organism of low intellect but strong physique - the wild man of today." He argued that they lacked "one or the other of the factors essential to a moral life - an understanding of right and wrong, and the power of control," and that these weaknesses made them strongly susceptible to becoming criminals, paupers, and prostitutes. Goddard was unsure whether mental deficiency resulted from the presence in the brain of something that inhibited normal development or from the absence of something that stimulated it. But whatever the cause, of one thing he had become virtually certain: it behaved like a Mendelian character. Feeblemindedness was "a condition of mind or brain which is transmitted as regularly and surely as colour of hair or eyes."[3]

Feeblemindedness was not only inherited; it was also said to be increasing at a socially menacing rate in both the United States and Canada. Between 1918 and 1922, a survey of mental deficiency in seven Canadian provinces was conducted under the auspices of the Canadian National Committee on Mental Hygiene (CNCMH). It found that in all seven the incidence of feeblemindedness was high and a threat to society on grounds that feeblemindedness was a primary cause of poverty, crime, and prostitution. In 1920, Helen MacMurchy, an energetic advocate of public health coupled to eugenics, published *The Almosts: A Study of the Feebleminded*. Addressed to laypeople, the book contended that feebleminded Canadians cost higher taxes because of their need for care and clogged the hospitals and reformatories. While they represented only three to five percent of the population, they accounted for half or more of alcoholics, juvenile delinquents, and unmarried mothers, not to mention between 29 percent and 97 percent of prostitutes.[4]

The backbone of the North American eugenics movement comprised people drawn from the white middle and upper middle classes, especially professional groups. Its supporters included prominent laymen and scientists, particularly geneticists, for whom the science of human biological improvement offered an avenue to public standing and usefulness. The eugenics leadership also included a number of medical practitioners, especially those who worked with people suffering from mental diseases and disorders. Its ranks were in addition composed of a significant number of women. By the cultural standards of the day, women were held - and held themselves - to be especially concerned with issues of child and family welfare. It was thus natural for them to find opportunity in the public sphere through eugenics, a movement that was heavily concerned with the bearing and development of children and with the impact of heredity on the family and, through the family, on society.

Much of eugenics, in fact, belonged to the wave of progressive social reform that swept through the United States and Canada during the early decades of the century. Women engaged in eugenics were often also involved in movements for the child labour reform, the improvement of nutrition and health, and the care of the mentally handicapped. For progressive reformers, eugenics was a branch of the drive for social perfection that many reformers of the day thought might be achieved through the deployment of science to good social ends. Eugenics, of course, also drew significant support from social conservatives, concerned to prevent the proliferation of lower-income groups and save on the cost of caring for them. The progressives and the conservatives found common ground in attributing phenomena such as crime, slums, prostitution, and alcoholism primarily to biology and in believing that biology might be used to eliminate these discordances of modern urban, industrial society.

Eugenics in North America was distinguished from its counterpart in Britain by its emphasis on race. By "race," eugenicists of the day did not mean primarily differences between blacks and whites. They meant differences between white, Anglo-Saxon or Nordic peoples and the immigrants flooding into North America during the period from Eastern and Southern Europe. Like eugenic scientists elsewhere, many American eugenicists held different national groups and

'Hebrews' to represent biologically different races and express different racial traits. Davenport found the Poles "independent and self-reliant though clannish"; the Italians tending to "crimes of personal violence"; and the Hebrews "intermediate between the slovenly Servians and the Greeks and the tidy Swedes, Germans, and Bohemians" and given to "thieving" though rarely to "personal violence." He expected that the "great influx of blood from South-eastern Europe "would rapidly make the American population "darker in pigmentation, smaller in stature, more mercurial ... more given to crimes of larceny, kidnapping, assault, murder, rape, and sex-immorality."[5]

Such observations were based upon crude, often anecdotal anthropological data, but IQ studies by Goddard and others had it that feeblemindedness occurred with disproportionately high frequency among lower-income and minority groups - notably recent immigrants in the United States from Eastern and Southern Europe. The seemingly deleterious impact of immigration acquired further authoritative backing after World War I, upon analysis of the IQ tests that had been administered to the thousands of draftees in the US Army.

The psychologist Robert Yerkes, the head of the testing program, and others claimed that the tests were almost entirely independent of the environmental history of the examinees, and that they measured 'native intelligence'; but the tests were biased in favour of scholastic skills, and test performance thus depended on the educational and cultural background of the person tested. A post-war testing vogue generated much data concerning the 'intelligence' of the American public, yet the volume of information was insignificant compared with that from the wartime test program, which formed the basis of numerous popular books and articles about intelligence tests and their social import. According to a number of popular analyses of this data, almost four hundred thousand draftees - close to one-quarter of the draft army - were unable to read a newspaper or to write letters home. Particularly striking, the average white draftee - and, by implication, the average white American - had the mental age of a thirteen-year-old.

The psychologist Carl Brigham, one of the wartime Army testers, extended the analysis of the Army data in 1923, in his book *A Study of American Intelligence*. The Army data, Brigham said, constituted "the first really significant contribution to the study of race differences in mental traits." Brigham found that according to their performance on the Army tests the Alpine and Mediterranean "races" were "intellectually inferior to the representatives of the Nordic race." He declared, in what became a commonplace of the popular literature on the subject, that the average intelligence of immigrants to the United States was declining.[6] The IQ test results reinforced the overall eugenic perception that 'racial degeneration' was occurring in the United States, and that a good deal of the trend was attributable to the immigrants flooding into the country from Eastern and Southern Europe.

Eugenicists did not concern themselves much, if at all, with blacks. To be sure, the IQ surveys indicated that the average intelligence of black Americans appeared to be just as low as most white Americans had long liked to think it. The Army test data, and various test surveys disclosed that blacks accounted for a disproportionately large fraction of the feebleminded; according to the Army test data, the average black person in the United States had the mental age of a ten-year-old. Blacks nevertheless did not foster eugenic anxieties, largely, it seems, because eugenicists did not count them as contributors to the quality of American civilisation. Or more important, one might say, as threats to that quality. The segregation of American society kept blacks isolated and under control. Eugenics in the Deep South illustrates the point: The large majority of blacks lived in the Southern region of the United States, but blacks were not objects of interest to southern eugenicists, where the social control of them was stringent. The South also had few recent immigrants. The overall aim of southern eugenicists was the preservation of the quality of the white population, and its target was the region's white, native "rubbish," in the phrase of an Atlanta paediatrician.[7]

American eugenicists fastened on British data which indicated that half of each succeeding generation was produced by no more than a quarter of its married predecessor, and that the prolific quarter was disproportionately located among the dregs of society. Before the war in the United States, leading eugenicists had warned that excessive breeding of the lower classes was giving the edge to the less fit. The growth of IQ testing after the war gave a quantitative authority to the eugenic notion of fitness: the vogue of mental testing not only encouraged fears regarding the "menace of the feeble-minded"; it also identified the source of heedless fecundity with low-IQ groups, especially immigrants, and it equated national deterioration with a decline in national intelligence.

Canadian eugenicists also identified the "menace of the feebleminded" partly with the immigration of "defective aliens." Drawing on the work of Goddard, MacMurchy estimated that Canada was admitting more than 1,000 feebleminded immigrants a year. Like analysts of immigrants in the United States, Canadian analysts held that mental defectiveness was disproportionately present among immigrants from Eastern and Southern Europe and that it was mostly inherited. The CNCMH survey typically found that the recent wave of Slavic immigrants to Alberta was marked by a high incidence of feeblemindedness. It was claimed that some 70 percent of patients in the mental hospitals of Alberta were foreign born, that there were more people in the mental hospitals of Canada than in all the general hospitals put together, and that hardworking tax payers were having to support these human drains on the public welfare.[8]

By permitting the immigration of mentally deficient aliens to continue unabated, Canada was said to be committing "race suicide." It was not only the absolute numbers of the immigrants that worried Canadian eugenicists; it was also that the newcomers seemed to proliferate to excess, bringing the threat of the differential birth rate to Canada. In Canada, as elsewhere in Anglo-American eugenic circles, the differential birth rate was often attributed to high sexual drive coupled with an irresponsibility that was thought to be inherent among immigrants - the same high degree of eroticism that was alleged to make many of them turn to prostitution. Whatever the cause of the differential birth rate, eugenic reasoning held that if immigrant deficiencies were hereditary and Eastern European immigrants outreproduced natives of Anglo stock, then inevitably the quality of the Canadian population would decline.[9]

* * *

A key platform plank of eugenicists in the United States was the restriction of immigration, which was achieved in an act of Congress in 1924. The measure severely restricted immigration from Eastern and Southern Europe. It had broad public support. It would have passed without the support of eugenicists, but eugenicists provided a biological rationale for the measure.

An Alabama eugenicist remarked early in the century that it was "essentially a state function" to restrain "the procreative powers" of the unfit. Eugenicists in North America offered the expertise available in eugenic research institutions to state and national governments for the formation of biologically sound public policy. They advised that the state should interfere in human propagation so as to increase the frequency of socially good genes in the population and decrease that of bad ones. The interference was to take two forms: One was 'positive' eugenics, which meant manipulating human heredity and/or breeding to produce superior people. The other was 'negative' eugenics, which meant improving the quality of the human race by eliminating or excluding biologically inferior people from the population.

Positive eugenic themes were certainly implied in the so-called 'Fitter Family' competitions that were a standard feature of the eugenic programs that were sponsored at a number of state fairs during the 1920s in the United States. These competitions were held in the 'human stock' sections of the fairs. At the 1924 Kansas Free Fair, winning families in the three categories - small, average, and large - were awarded a Governor's Fitter Family Trophy, which was presented by Governor Jonathan Davis, and "Grade A Individuals" received a medal that portrayed two diaphanously garbed parents, their arms outstretched toward their (presumably) eugenically meritorious infant. It is hard to know what made these families and individuals stand out as fit, but some evidence is supplied by the fact that all entrants had to take an IQ test - and the Wasserman test for syphilis.

Much more was urged for negative eugenics, notably the passage of eugenic sterilisation laws. In the United States by the late 1920s, some two dozen American states had framed compulsory eugenic sterilisation laws, often with the help of the Eugenics Record Office, and enacted them. The laws were declared constitutional in the 1927 U.S. Supreme Court decision in the case of *Buck v. Bell*, in which Justice Oliver Wendell Holmes delivered himself of the opinion that three generations of imbeciles were enough.[10]

Eugenic sterilisation was not uniformly adopted in the United States. More than a third of the states of the union declined to pass sterilisation laws, and most of those that did pass them did not enforce them. In regional terms, relatively few states in the Northeastern United States passed these laws. Only three states in the Old South did. In the Northeast and to some degree elsewhere, including Louisiana, the passage of sterilisation measures was effectively resisted by Roman Catholics.

Catholics strongly opposed sterilisation, partly because it was contrary to Church doctrine, partly because a very high fraction of recent immigrants to the United States were Catholics and were thus disproportionately placed in jeopardy of the knife. Passage was accomplished largely in the Middle Atlantic States, the Midwest, and in California, the champion of them all. As of 1933, California had subjected more people to eugenic sterilisation than had all other states of the union combined. Wherever they were passed the laws reached only to the inmates of state institutions for the mentally handicapped or mentally ill. People in private care or in the care of their families eluded them. They thus tended to work discriminatorily against lower-income and minority groups. California, for example, sterilised blacks and foreign immigrants at nearly twice the per-capita rate as the general population.[11]

Like their counterparts in the United States, Canadian eugenicists also agitated for immigration restriction and sterilisation of the mentally deficient. The CNCMH argued for guarding the gates against the insane and the feebleminded, and in 1924, the convention of the United Farm Women of Alberta established a committee to seek to prohibit entry into Canada of immigrants who were feebleminded, epileptic, tubercular, dumb, blind, illiterate, criminal, and anarchistic. Apparently, the movement for immigration restriction failed. But the drive for sterilisation was another story. Agitation for sterilisation began before World War I, with its advocates explaining that segregation of feebleminded people was insufficient; destruction of their capacity to reproduce would be far more economical, since then they would no longer have to be housed at state expense. Helen MacMurchy contended, "We must not permit the feeble-minded to be mothers of the next generation." A bill to authorise sterilisation of the mentally deficient was introduced into the Ontario legislature in 1912 but failed.[12]

Resistance to sterilisation laws was strong in the eastern provinces of Canada, not least because Catholics there were politically powerful. But in the Western provinces, which had many fewer Catholics, support for sterilisation grew after World War I. A look at the Alberta case illuminates the forces at work. Sterilisation acquired increasing advocacy from highly respected members of society like Emily Murphy, a suffragist, pioneer as a female police magistrate in Alberta, and member of the board of visitors to Alberta's mental institutions. In addresses to women's organisations, she spoke energetically in favour of sterilisation of the feebleminded, declaring that 75 percent of the cause of feeblemindedness and insanity was heredity, warning of the threat that the differential birth rate posed, and quoting Henry Goddard to the effect that "every feebleminded person is a potential criminal." In 1924, the results of the CNCMH survey, together with an initiative from the United Farmers of Alberta (UFA), prompted the first introduction of a measure in the Alberta legislature to sterilise mental patients, and in 1925, the UFA endorsed compulsory sterilisation of the mentally deficient. President Margaret Gunn, of the United Farm Women of Alberta argued that the procreation of derelicts who would "lower the vitality of our civilization" had to be prevented. Responding to opposition that sterilisation would violate the civil liberties of inmates, she declared that "democracy was never intended for degenerates."[13]

The pro-sterilisation forces received a substantial boost from a Royal Commission on Mental Hygiene that was established under the chairmanship of Dr E J Rothwell in 1925 and delivered a preliminary report in 1927 and a final report in March 1928. Intended to examine the issue of mental deficiency in British Columbia, it nevertheless appears to have played an important role in advancing the issue in Alberta. Its views were authoritative, since the Commission obtained evidence from experts in Eastern Canada and in the United

States. It concluded that the growth in the frequency of mental deficiency had been exaggerated, but it did find that facilities for care of people with mental diseases and disorders were increasingly overcrowded. It was apparently divided on what causes mental problems, with some of its witnesses having argued that they were fundamentally psychodynamic while others held that they were mainly hereditary. Nevertheless, the principal remedy it advanced was sterilisation, which it argued for as an economic necessity for coping with rising institutional costs. Witnesses from California testified that sterilisation programs were highly successful there. With sterilisation, patients could enjoy the greater liberty of living relatively normal lives in the community. Proponents of the operation claimed that sterilisation was morally beneficial for the mentally deficient, fostering greater order and self-control in their lives. The Commission thus endorsed sterilisation for people in mental institutions who consented to the procedure and who "might safely be recommended for parole from the institution and trial return to community life, if the danger of procreation with its attendant risk of multiplication of the evil by transmission of the disability to progeny were eliminated."[14]

On March 25, 1927, George Hoadley, the Minister of Health in the United Farmers of Alberta Government, introduced a bill that would authorise the sterilisation of people who suffered from mental deficiency or disorders and who resided in state institutions for their care.[15] The bill provoked a long and bitter debate. The measure was denounced as a violation of the patient's civil liberties. It was attacked on grounds that the coupling of discharge from the institution to consent to undergo the operation made a mockery of the idea of freely given consent; Laudas Jolly, a UFA member from St Paul, noted that the measure offered "mutilation as the price of liberty for inmates of mental hospitals." The bill was also opposed as without scientific foundation and as an offence to moral and religious principles. Nevertheless, the bill had broad support from medical practitioners and laypeople alike, including the UFA and the United Farm Women of Alberta, the women's section of the Dominion Labor Party in Calgary, the Canadian Mental Hygiene Society, the Women's Christian Temperance Union, and the College of Physicians and Surgeons. Premier Brownledd called sterilisation far more effective than segregation and insisted that "the argument of freedom or right of the individual can no longer hold good where the welfare of the state and society is concerned." In March 1928, the measure passed by a solid majority of 34-11.[16]

During the 1930s, the economic depression strengthened support for eugenic sterilisation in Canada and the United States; largely, one is inclined to think, so that the state homes for the mentally handicapped could release more inmates and thus save money. Madge Thurlow Macklin, a geneticist at the University of Western Ontario, an organiser of the Eugenics Society of Canada, and an outspoken advocate of eugenic sterilisation of the feebleminded, warned against the differential birth rate, raising fears that Canadian society, including its public schools, was being swamped by people with mental deficiency. She declared, "We care for the mentally deficient by means of taxes, which have to be paid for by the mentally efficient ...". She insisted that sterilisation was warranted on grounds of "incontrovertible scientific facts." In 1937, Macklin visited Germany, surveyed the Nazi programs for the mentally ill, and returned to Canada with her support for sterilisation undiminished.[17]

In the United States and Canada during the 1930s, sterilisation rates climbed. In 1930, a Eugenics Society of Canada was founded, its membership heavy with medical doctors, and the treasurer of the Eugenics Society of Canada, in a publication called *Sterilization Notes*, pointed to the "successes" of the sterilisation programs already under way in California and Germany as well as in Alberta. He stressed sterilisation as a mean of reducing the burdens of relief, and an official of the Edmonton Public School Board held that further sterilisation of defectives would save considerable money in the costs of crime and unemployment.[18] Perhaps not surprisingly, Eastern European immigrants and the Métis Indians were sterilised under the Alberta program at a far higher rate than their proportion in the population would have warranted.[19]

* * *

Paradoxically, even while sterilisation rates were rising, opinion was turning increasingly against eugenics, not least because of its association with the Nazis. In Alabama, for example, attempts to pass a sterilisation law in the mid-1930s prompted a Methodist newspaper to warn that the "proposed sterilisation bill is a step" toward the "totalitarianism in Germany today." There the "state is taking private matters - matters of individual conscience, and matters of family control - in hand, and sometimes it's a rough hand, and always it's a strong hand." Governor Bibb Graves put the issue more succinctly: "The great rank and file of the country people of Alabama do not want this law; they do not want Alabama, as they term it, Hitlerized."[20]

Scientific opinion had started turning against eugenic doctrine in the 1920s because of the shoddiness that coloured its theories of human heredity, and by the 1930s the scientific critique was growing increasingly forceful and convincing. Psychologists held that the diagnosis of mental deficiency depended too heavily upon the results of intelligence tests. Mental-health professionals learned from experience that a number of people committed to institutions as feebleminded on the basis of the Binet-Simon tests were capable of leading successful independent lives. One might be slow at lessons but possess more than adequate common sense and be a useful member of society. By the late nineteen-twenties, Henry H. Goddard himself had, as he said, gone over "to the enemy," conceding that only a small percentage of the people who tested at mental ages of twelve or less were incapable of handling their affairs with ordinary prudence and competence. By the 1930s, the growing consensus of scientific opinion was that the alleged menace of the feebleminded was a myth, a speculation totally without foundation.

Conclusions about the inheritance of mental deficiency were undercut by the fact that the children of men and women admitted to asylums often did not themselves appear to be similarly afflicted. Some deficiencies were in fact inherited, but matings between mentally deficient people did not necessarily produce deficient offspring in the numbers predicted by Mendel's laws as eugenicists used them. In the speculation of geneticists, the reason was that many traits were polygenic in origin - that is, the result of many genes. Then, too, the mental deficiency suffered by one parent might originate in a different set of genes from that found in another. In sum, by the 1930s just what genetic combinations made for mental deficiency were, to say the least, unclear. Mental deficiency was found in many forms. Complex in its expression, it was presumably diverse in its causes.

Science aside, after World War II, eugenic sterilisation also became offensive to moral sensibilities in most regions of the Western World because of its association, now revealed, with the Nazi death camps. The Eugenics Society of Canada died around the end of World War II. Sterilisations continued in several American states through the early 1960s, and in Alberta until 1972, but eugenics had become a dirty word in North America.

Yet even as eugenics fell completely out of fashion, genetic research was raising the curtain on a new, potentially revolutionary era in the control of heredity, including the human variety. Rapid progress in human cytogenetics - particularly the recognition in 1959 that Down's syndrome arises from a chromosomal anomaly - soon made prenatal diagnosis possible, with the option of abortion for women at risk of giving birth to children with severe chromosomal disorders. The unveiling of the structure of DNA, in 1953, opened the door to the discovery of how genes actually control the development (and misdevelopment) of organisms. By the mid-1960s, it was understood that genes embody a code, written into their chemical structure, that instructs the cell what specific proteins to manufacture. Finding proteins associated with diseases made it possible to identify flaws in DNA that generated illness and to detect disease genes in recessive carriers and foetuses homozygous for illnesses such as Tay-Sachs disease and sickle-cell anaemia. The working out of the genetic code inspired neo-Galtonian visions. As early as 1969, Robert Sinsheimer, a prominent molecular biologist at the California Institute of Technology declared that "for the first time in all time, a living creature understands its origin and can undertake to design its future" - and, in consequence, might eventually control its own evolution.[21]

Yet the history of eugenics in North America strongly argues that utopian genetic visions merit scepticism. The most enthusiastic advocates of eugenics in North America during the first half of the century included a significant portion of social progressives. They thought that the merger of science with demographic need would lead to social melioration. They passed sterilisation laws that were upheld by the U.S. Supreme Court, and the court's decision was cited in Alberta as authority for its own sterilisation measure.

The case of *Buck v. Bell* had originated in Virginia. In the course of hearing it, the Virginia Board was presented with evidence that Carrie Buck, the patient who was proposed for sterilisation, was feebleminded and that her feeblemindedness was hereditary in the Buck line. The gathering of this evidence satisfied the requirements of the law. It also satisfied the requirements of the U.S. Constitution, according to Justice Oliver Wendell Holmes, who observed in his opinion for the majority: "There can be no doubt that so far as procedure is concerned the rights of the patient are most carefully considered, and as every step in this case was taken in scrupulous compliance with the statute after months of observation, there is not doubt that in that respect the plaintiff in error has had due process of law."[22]

Virginia was not alone in giving attention to at least the form of protecting individual rights in its sterilisation law. In 1932, a review of sterilisation laws in other states concluded that they all provided for reasonable notice to the proposed sterilizee and provided that person an opportunity for self-defence. In the mid-1930s, when 28 states had eugenic sterilisation laws on their books, a student (and enthusiast) of them noted: "There is a growing tendency also for the statutes to define more specifically the criteria by which the courts shall decide that the particular individual falls within, or without, the sterilisation category. The qualities of the individual, his own case history, and where possible to secure them, the description of natural qualities of his nearest blood-kin, are essential. Provisions for hearing both sides of the case, and ample provisions for appeal to higher courts are made so that there is continuously less danger that the State will make unjust decisions in the proposed sterilisation of a particular defective individual."[23]

In retrospect, it is evident that what was legal and constitutional left a good deal to be desired measured against standards of human rights. The scientific evidence concerning Carrie Buck's feeblemindedness was flimsy and would not have stood close scrutiny by a scrupulous geneticist or psychologist even in the 1920s. The American sterilisation laws were drawn in form to protect the rights of individuals, but in substance and practice they failed to do so. The Alberta law provided no such protection even in form after it was modified in 1937 to eliminate the requirement that the patient or the patient's guardian give consent for the procedure. The record of the experiment in eugenic sterilisation in North America offers a powerful indication that the uses of genetic science and genetic information today warrant considerable care and attention not only in law but in practice to civil liberties, individual rights, and social decency.

References:

[1] Francis Galton, *Inquiries into the Human Faculty* (London: Macmillan, 1883), pp. 24-25; Karl Pearson, *The Life, Letters, and Labours of Francis Galton* (3 vols. in 4; Cambridge: Cambridge University Press, 1914-1930), IIIA, 348.

[2] Daniel J Kevles, *In the Name of Eugenics: Genetics and the Uses of Human Heredity* (Cambridge, MA: Harvard University Press, 1995), p. 62.

[3] Henry H Goddard, *Feeble-mindedness: Its Causes and Consequences* (New York: Macmillan, 1914), pp. 4, 7-9, 14, 17-19, 413, 504, 508-9, 514, 547.

[4] Angus McLaren, *Our Own Master Race: Eugenics in Canada, 1885-1945* (Toronto: McCleland & Stewart, Inc., 1990), pp. 25-41, 93.

[5] Charles B Davenport, *Heredity in Relation to Eugenics* (New York: Henry Holt, 1911),pp. 216, 218-19, 221-22.

[6] Kevles, *In the Name of Eugenics*, pp. 82-83.

[7] Edward J Larson, *Sex, Race, and Science: Eugenics in the Deep South* (Baltimore: Johns Hopkins University Press, 1995), pp. 1, 9, 93.

[8] McLaren, *Our Own Master Race*, pp. 50-51, 99; Tim Christian, "The Mentally Ill and Human Rights in Alberta: A Study of the Alberta Sexual Sterilization Act" (n.d., Faculty of Law, University of Alberta), pp. 14-15.

[9] McLaren, *Our Own Master Race*, pp. 50-51, 72.

[10] *Buck v. Bell*, 274 U.S., 201-207.

[11] Larson, *Sex, Race, and Science*, pp 37-38.

[12] McLaren, *Our Own Master Race*, pp. 38-43, 64; Christian, "The Mentally Ill and Human Rights in Alberta," pp. 10-12.

[13] Christian, "The Mentally Ill and Human Rights in Alberta," pp. 8-12.

[14] McLaren, *Our Own Master Race*, pp. 96-98.

[15] Christian, "The Mentally Ill and Human Rights in Alberta", pp. 20-22; "An Act respecting Sexual Sterilization," 1928, Chapter 37.

[16] Christian, "The Mentally Ill and Human Rights in Alberta", pp. 2,13, 15, 20-22.

[17] McLaren, *Our Own Master Race*, pp. 136-47.

[18] Christian, "The Mentally Ill and Human Rights in Alberta," p. 27; McLaren, *Our Own Master Race*, pp. 114-121, 157.

[19] Christian, "The Mentally Ill and Human Rights in Alberta," pp. 76, 81, 85, 90, 118-21.

[20] Larson, Sex, Race, and Science, p. 146.

[21] Robert Sinsheimer, "The Prospect of Designed Genetic Change," *Engineering and Science*, 32(April 1969), p.8.

[22] Buck v. Bell.

[23] Harry H Laughlin, "Further Studies on the Historical and Legal Development of Eugenical Sterilization in the United States", *American Association on Mental Deficiency, Proceedings*, XLI (May 1-4, 1936), 100; Jacob H. Landman, *Human Sterilization: The History of the Sexual Sterilization Movement* (New York: Macmillan, 1932), pp. 16-17.

Index

A

abortion, 189, 191, 223
Abortion Law Reform Association, 41
Alliance Nationale contre la Dépopulation, 182
American Civil War, 13
American Society of Human Genetics, 123
Ammon, Otto, 179
Annals of Eugenics, x, 171
Annals of Human Genetics, 171
Annan, Noel, 5, 6, 13, 17
Anthropology, 2
Anti-Corn Law League, 28
Association of British Insurers, 121, 125, 127
Association of Public Health, Swedish, 193

B

Bagehot, W, 4
Baker, John, 68, 70, 71
Bartlett, V, 132
Belfast, 14
Belfast's Natural History and Philosophical Society, 14
Berger, Gaston, 188
Beveridge, Sir William, 86, 87, 91, 95, 110
Binet, Alfred, 128, 129, 153
Binet-Sanglé, Dr, 183, 204
Biometrics, 156
Biometrics Bulletin, 156
Biometrika, 156
Birmingham Heredity Society, 29
birth control, 27, 53, 54, 57, 60, 61, 62, 63, 64, 70, 73, 75
Birth Control Investigation Committee, 43, 47, 54, 63, 64, 68, 70, 72, 77, 78
Blacker, C P, x, xi, xiii, 28, 43, 44, 46, 47, 82, 84, 89, 91, 92, 93, 98, 99, 100, 101, 102, 103, 107, 109, 110
Boer War, 14
Borgstôm, C A, 197
Bourgeois, Jean, 188, 189, 205
Brigham, Carl, 213
British Library, ii
British Social Hygiene Council, 68
Broberg, Gunner, 192, 193, 206
Broca, Paul, 177, 204
Browne, Stella, 40, 41, 49, 50
Buckle, Henry, 4
Burt, Cyril S, xiii, 133, 134, 135, 138, 142, 144, 145, 147, 148, 153, 154
Butler family, 6

C

Cadbury, L J, 103
Cambridge, University of, 2, 132
Canadian Mental Hygiene Society, 220
Canadian National Committee on Mental Hygiene, 210, 214, 217, 218
Candolle, Alphonse, 4, 5
Caron, Alfred Charles, 177, 180
Carrel Foundation, 187, 189

Carrel, Alexis, 173, 176, 184, 186, 187, 189, 190
Carroll, J B, 130, 135, 136, 153
Carr-Saunders, A M, 72, 79, 80, 84, 85, 86, 89, 91, 93, 98, 99, 100, 101, 102, 103, 107, 108, 109
Cattell, Raymond B, xiii, 132, 133, 134, 135, 136, 137, 138, 139, 141, 145, 150, 151, 153, 154
Chance, Clinton, 66
Charity Organization Society, 15, 30, 33
China, People's Republic of, 122
Churchill, Winston, 25, 26
Clarion, The, 41
Close, Sir Charles, 103
Cobb, J A, 159, 160, 161, 162, 166, 170
Cobden, Richard, 28
College of Physicians and Surgeons, 220
Consultative Eugenics Committee of Norway, 193
Crew, F A E, 84, 85, 86, 91, 93
Crichton-Browne, Sir James, 22

D

Darwin family, 6
Darwin, Charles, viii, 9, 10, 11, 12, 17, 18, 19, 22, 114, 115
Darwin, Leonard, x, 22, 23, 27, 28, 30, 31, 32, 158, 167
Davenport, Charles B, 209, 212, 225
differential birth-rate, 25
differential fertility, 11, 56
Dominion Labor Party, 220
Douglas, J W B, 105

Drysdale, George, 182, 204

E

Ecole d'Anthropologie, 186
Edwards, J H, 172
Ellis, Havelock, 57, 77
Eugenics Education Society, 3, 14, 15, 16, 20, 21, 22, 23, 24, 25, 28, 29, 30, 31, 32, 33, 53, 56, 57, 58, 59, 60, 62, 77, 78, 83, 112, 117, 118, 126
Eugenics Laboratory, 20, 22, 23
Eugenics Records Office, 209
Eugenics Review, xiii, 24, 26, 31, 35, 57, 61, 74, 75, 77, 78, 79, 80, 83, 85, 89, 94, 96, 98, 110, 157, 159, 161, 162, 164, 165, 166, 167
Eugenics Society, 52, 54, 55, 61, 62, 63, 64, 67, 68, 69, 70, 71, 72, 73, 74, 75, 77, 78, 79, 80, 81, 82, 83, 85, 90, 94, 95, 97, 98, 99, 100, 101, 104, 106, 108, 110
Eugenics Society of Canada, 220, 222
Evang, Karl, 197
Eyre, Governor, 13
Eysenck, Hans, 133, 134, 138, 139, 140, 141, 142, 154

F

Fabian Society, 26, 29, 33
factor analysis, 129, 131
family allowances, 41, 46
Family Planning Association, xi, 48, 51, 68, 70, 74
Federley, Harry, 195, 207
feeble-minded, 29
Fisher, R A, xiii, 2, 16, 18, 145, 156, 157, 158, 159, 160, 161,

162, 163, 164, 165, 166, 167, 168, 169, 170, 171, 172
Foundation for the Study of Human Problems, 173
François Perroux Department of Biosociology, 188
Freewoman, The, 41
French Academy of Medicine, 179
French Foundation for the Study of Human Problems, 187, 188, 189
French League for Human Rights, 185
French National Committee on Ethics, 191
French Society of Eugenics, 186

G

g, 129, 130, 131, 135
Gall, F J, 8
Galton Institute, i, ii, v
Galton Laboratory, 68
Galton Lecture, xiii, 69, 95, 98, 99, 165, 171, 172
Galton Lecture of 1930, 67
Galton, Francis, viii, ix, x, xi, xiii, xiv, 1, 2, 3, 4, 5, 6, 7, 8, 9, 10, 11, 12, 13, 16, 17, 18, 19, 20, 21, 22, 23, 24, 32, 33, 34, 52, 53, 56, 63, 65, 69, 75, 77, 78, 80, 113, 114, 115, 116, 128, 142, 156, 157, 159, 163, 170, 171, 172, 173, 176, 177, 178, 202, 207, 208, 225
Garrod, Archibold, 115, 116, 126
genetics, 53, 60, 62, 65, 175, 183, 190, 193, 198, 207
Gessain, Robert, 188

Glass, David V, 89, 91, 98, 99, 100, 101, 102, 103, 104, 106, 107, 109, 110
Goddard, Henry H, 209, 210, 212, 214, 218, 221, 225
Gotto, Sybil, x, 22
Grebenik, E, 101, 104
Gros, André, 188
Guesde, Jules, 178
Guilford, J P, 136, 137, 138, 139, 154
Gunn, Margaret, 218
Günther, Hans F K, 179

H

Haeckel, Ernst Heinrich, 177
Haldane, J B S, 2, 157, 168, 169, 170, 171, 175, 197
Hamilton, W D, 159, 162, 164, 168
Hoadley, George, 219
Hofsten, Nils von, 197
Hogben, Lancelot, 66, 168, 169, 170, 171, 172
Holmes, Oliver Wendell, 216, 223
Horder, Lord, 84, 93, 99, 100, 103, 109
Hubback, Eva, 44, 46, 50
human genetics, 198
Human Genetics Advisory Committee, 121
Human Genome Project, 119, 120, 125, 126, 170
Huxley, Julian, 61, 63, 65, 78, 79, 80, 175, 191, 197
Huxley, Thomas Henry, 10, 18
Huxley's 1936 Galton Lecture, 65

I

inclusive fitness, 162, 164, 166, 168
Institute Alfred Fournier, 185
Institute for Racial Biology, 193, 203
International Biometric Society, 156, 157
International Genetics Federation, 123
Internationale, 181
IQ, 128, 129, 148, 154

J

Johannsen, Wilhelm, 195, 207
Joint Committee on Voluntary Sterilization, 68
Jones, Ernest, xiii

K

Kemp, Tage, 195, 197, 198, 207
Kenealy, Arabella, 38, 39, 40, 49
Keynes, John Maynard, 69
Kuczynski, R R, 84, 100, 102, 104

L

labour parties, 198
Lamarckism, 179
Landry, Adolphe, 184
Lapouge, Georges de, 176, 177, 178, 179, 203, 204
Lavater, J K, 8
Layton, Sir Walter, 100, 103
League for Human Regeneration, 181, 185
Lenz, Fritz, 198
Lewis-Faning, E, 101, 104, 107
Leybourne, G, 102
Lindsay, James Alexander, 14
linkage theory, 171

Lloyd George, David, 25, 26
London School, 130, 131, 132, 133, 135, 136, 141
London School of Economics, 15, 19, 87, 97, 100, 101, 102, 106
Lucas, Prosper, 177
Lundborg Herman, 177, 179, 193, 207
Lutherism, 198

M

MacMurchy, Helen, 210, 214, 217
Mallet, Sir Bernard, 54, 64, 77, 78, 83, 84, 85, 86, 88, 89, 91, 93, 96
Malthusian theories, 10
Malthusianism, 10
Manifesto of Geneticists, 175
March, Lucien, 184
Marriage Guidance Council, 68
Martial, Dr, 183
Marxist, 178
Mayr, Ernst, 191
Medical Women's Federation, 37, 42, 49
Mendel, Gregor, xii, 24, 113, 114, 115, 126, 175, 208, 222
Mendelism, 2, 16
Mental Deficiency Acts, 33
Merlet, Jean, 188
Mill, John Stuart, 4, 10, 13
Mother's Aid Agencies, 197
Muller, Hermann S, 175
Muller, J H, 191, 197
Murphy, Emily, 218
Myrdal, A & G, 194

N

Nachtsheim, H, 197

National Birth Control Association, 44, 47, 48, 68, 70, 72, 73, 74, 80
National Birth Control Council, 43
National Conference of Labour Women, 44
National Council of Women, 44
National Institute for Demographic Studies, 188, 205, 206
National Insurance Act, 25
National Society for Women's Service, 42
National Union of Societies for Equal Citizenship, 42, 44, 50
National Union of Women's Suffrage Societies, 42
Nazism, 70, 71, 123, 147, 173, 175, 185, 220, 221, 222
neo-Lamarckism, 183
neo-Malthusian, xi, 176, 181, 185
neo-Malthusianism, 181, 189
neo-Malthusians, 176, 181, 182, 183
New Feminism, 41
New Feminist, 46
New Woman, x
Nilsson-Ehle, Herman, 195, 207

O

Osborne, Reverend Henry, 15

P

Painter, T S, 118
parental expenditure, 159, 160, 161, 162, 166
Pasteur, Louis, 184
Pearl, Raymond, 84, 86, 87, 88
Pearson, Karl, 1, 2, 9, 10, 16, 17, 18, 22, 24, 56, 57, 77, 156, 167
pedigree collection, 115, 117
pedigree studies, 118
Penrose, L S, 2
People's Budget, 25
Pinard, Adolphe, 177, 179, 184, 189
Pittard, Eugène, 183, 205
Plato, 3, 153
Plauzoles, Sicard de, 184, 185
Plomin, Robert, 146
Political and Economic Planning, 68, 69, 72
political economy, ix
Population, 88, 109
Population (Statistics) Act 1938, 102
Population Investigation Committee, 68, 69, 72, 79
Population Studies, xii, 82, 106, 109
Professional Classes War Relief Council, 32, 58
Prospective, 188

Q

Quetelet, L A J, 8

R

Reform Act of 1867, 13
reproductive value, 165, 166
Richet, Charles, 176, 183, 184, 203, 204
Robin, Paul, 176, 181, 182, 204
Rolleston, Sir Humphrey, 70, 79
Roll-Hansen, Nils, 192, 193, 206
Rostand, Jean, 176
Rowntree, Griselda, 105
Royal Geographical Society, 53, 87, 111

S

Sanger, Margaret, 81, 83, 84, 85, 96, 111
Sauvy, Alfred, 184, 205, 206
Schopenhauer, A, 9
Schreiner, Olive, x
Scott, Conway, 14
Scottish Council for Research in Education, 105
sex ratio, 160, 161, 162
Shaw, George Bernard, 23, 27
Sinsheimer, Robert, 223, 226
Slaughter, J W, 15
Smiles, Samuel, viii
Smith, C A B, 157
social Darwinism, 26, 175, 208
Society for the Promotion of Birth Control Clinics, 68
Spearman, Charles Edward, xii, 129, 130, 131, 132, 133, 134, 136, 154
sperm donors, 191
Stanford-Binet test, 129
Steincke, K K, 194
Stephen, FitzJames, 13
Stephen, Leslie, 13
sterilisation, 39, 40, 41, 44, 45, 46, 146, 174, 176, 182, 183, 192, 193, 194, 195, 196, 197, 201, 216, 217, 218, 219, 220, 221, 222, 223, 224
sterilisation laws, 195
Stoetzel, Jean, 188, 189
Stopes, Marie Carmichael, 39, 40, 60, 78
Sutter, Jean, 188, 205, 206
Swedish Society for Racial Hygiene, 193

T

Terman, Lewis Madison, 129, 155
The French Society of Eugenics, 185
Thomson, Godfrey, 148
Thurstone, L L, 134, 136, 155
Titmuss, Richard, 74
Topinard, 177
Toulouse, Edouard, 184, 190, 206
Trinity College, Dublin, 2
Twitchin bequest, xi, 67
Twitchin, Henry, 64
Tylor, E B, 4

U

University College, London, ix, 1, 2, 129, 132, 152, 157, 170
Urwick, E J, 16

V

venereal disease, 29
Venn, John, 2
Vernon, P E, 144, 148, 150, 155
Verschuere, Otmar von, 198
Vincent, Paul, 188
Volpar, 70, 79

W

Wallace, Alfred Russell, 114
Webb, Beatrice, 8, 11, 18, 19
Wedgwood family, 12
Wedgwood, family, 6
Weldon, W F R, 156, 157
welfare state, 174, 192, 195, 198
Welfare State, 206
welfare state eugenics, 192
Women's Co-operative Guild, 44, 59
Women's Institutes, 44

World Population Conference, 81, 83, 93, 111

Y

Yerkes, Robert, 212
Young, Michael, xii

Z

Zuckerman, Solly, 70

ALSO AVAILABLE IN THIS SERIES

MARIE STOPES, EUGENICS AND THE ENGLISH BIRTH CONTROL MOVEMENT

EDITED BY ROBERT A PEEL

Proceedings of the 1996 Conference of the Galton Institute

CONTENTS

Notes on the Contributors

Editor's Preface
Robert Peel

Introduction
John Peel

The Evolution of Marie Stopes
June Rose

Marie Stopes and her Correspondents: Personalising population decline in an era of demographic change
Lesley Hall

The Galton Lecture: "Marie Stopes, Eugenics and the Birth Control Movement"
Richard Soloway

Marie Stopes and the Mothers' Clinics
Deborah Cohen

"Marie Stopes: Secret Life" – A Comment
John Timson

Marie Stopes International Today
Patricia Hindmarsh

Index

ISBN 0950406627

Available post paid from the Institute's General Secretary Price £5.00